Multimedia Services in Wireless Internet

Wiley Series on Wireless Communications and Mobile Computing

Series Editors: Dr Xuemin (Sherman) Shen, *University of Waterloo, Canada*
                  Dr Yi Pan, *Georgia State University, USA*

The "Wiley Series on Wireless Communications and Mobile Computing" is a series of comprehensive, practical and timely books on wireless communication and network systems. The series focuses on topics ranging from wireless communication and coding theory to wireless applications and pervasive computing. The books offer engineers and other technical professionals, researchers, educators, and advanced students in these fields with invaluable insight into the latest developments and cutting-edge research.

Other titles in the series:

Misic and Misic: *Wireless Personal Area Networks: Performance, Interconnections and Security with IEEE 802.15.4*, January 2008, 978-0-470-51847-2

Takagi and Walke: *Spectrum Requirement Planning in Wireless Communications: Model and Methodology for IMT-Advanced*, April 2008, 978-0-470-98647-9

Perez-Fontan and Espiñeira: *Modeling the Wireless Propagation Channel: A Simulation Approach with MATLAB®*, April 2008, 978-0-470-72785-0

Ippolito: *Satellite Communications Systems Engineering: Atmospheric Effects, Satellite Link Design and System Performance*, August 2008, 978-0-470-72527-6

Lin and Sou: *Charging for Mobile All-IP Telecommunications*, September 2008, 978-0-470-77565-3

Myung and Goodman: *Single Carrier FDMA: A New Air Interface for Long Term Evolution*, October 2008, 978-0-470-72449-1

Wang, Kondi, Luthra and Ci: *4G Wireless Video Communications*, April 2009, 978-0-470-77307-9

Hart, Tao and Zhou: *Mobile Multi-hop WiMAX: From Protocol to Performance*, October 2009, 978-0-470-99399-6

Stojmenovic: *Wireless Sensor and Actuator Networks: Algorithms and Protocols for Scalable Coordination and Data Communication*, December 2009, 978-0-470-17082-3

Qian, Muller and Chen: *Security in Wireless Networks and Systems*, January 2010, 978-0-470-512128

Multimedia Services in Wireless Internet

Modeling and Analysis

Lin Cai
University of Victoria, Canada

Xuemin (Sherman) Shen
University of Waterloo, Canada

Jon W. Mark
University of Waterloo, Canada

A John Wiley and Sons, Ltd, Publication

Registered office
John Wiley & Sons Ltd, The Atrium, Southern Gate, Chichester, West Sussex, PO19 8SQ, United
Kingdom

For details of our global editorial offices, for customer services and for information about how to apply
for permission to reuse the copyright material in this book please see our website at www.wiley.com.

Library of Congress Cataloging-in-Publication Data

Cai, Lin, 1973-
    Multimedia services in wireless internet : modeling and analysis / by Lin Cai, Xuemin
(Sherman) Shen, and Jon W. Mark.
        p. cm.
    Includes bibliographical references and index.
    ISBN 978-0-470-77065-8 (cloth)
1. Wireless internet–Mathematical models. 2. Multimedia communications–Simulation
methods. 3. Wireless communication systems–Quality control. I. Shen, X. (Xuemin), 1958-
II. Mark, Jon W. III. Title.
    TK5103.4885.C35 2009
    621.382'1–dc22

                                                                                          2009005785

A catalogue record for this book is available from the British Library.

ISBN 9780470770658 (H/B)

Set in 11/13 pt Times by Sunrise Setting Ltd, Torquay, UK.
Printed in Great Britain by CPI Antony Rowe, Chippenham, Wiltshire.

Contents

About the Series Editors   xi

About the Authors   xiii

Preface   xv

1 Introduction   **1**
    1.1   Convergence of Wireless Systems and the Internet   1
    1.2   Main Challenges in Supporting Multimedia Services   3
    1.3   Organization of the Text   7

2 Packet-level Wireless Channel Model   **9**
    2.1   Introduction   9
    2.2   Finite-state Markov Model for Fast Fading Channels   10
       2.2.1   Finite-state Markov model for Rayleigh fading channels   . .   10
       2.2.2   Mapping finite-state Markov chain to two-state Markov chain   12
    2.3   Channel Model for Frequency-selective Fading Wireless Channels   .   13
       2.3.1   The Nakagami-m fading model   15
       2.3.2   LCR in the frequency domain   16
       2.3.3   FSMC in the frequency domain   17
       2.3.4   FSMC model for OFDM systems   19
       2.3.5   Simulation study   20
    2.4   Channel Model for Indoor UWB Wireless Channels with Shadowing   24
       2.4.1   The angular power spectral density of UWB signals   26
       2.4.2   People-shadowing effect on UWB channels   31
       2.4.3   Performance of MB-OFDM link with the people-shadowing
             effect   33
       2.4.4   Markov model for UWB channel with people shadowing   . .   36
       2.4.5   Numerical results   38
    2.5   Summary   40
    2.6   Problems   42

3   Multimedia Traffic Model   **45**
    3.1   Modeling VoIP Traffic   45
          3.1.1   VoIP 45
          3.1.2   VoIP traffic model   47
    3.2   Modeling Video Traffic   49
          3.2.1   Mini-source video model   49
          3.2.2   A simple, two-level Markovian traffic model   51
    3.3   Performance Study of Video over Wired and Wireless Links 55
          3.3.1   Transmission over a wired link   58
          3.3.2   Transmission over a wireless link   59
          3.3.3   Multiplexing heterogeneous traffic with class-based
                  queueing   60
          3.3.4   Simulation results   63
    3.4   Scalable Source Coding   70
    3.5   Summary   71
    3.6   Problems   71

4   AIMD Congestion Control   **75**
    4.1   Introduction   75
    4.2   AIMD Protocol Overview   77
          4.2.1   Acknowledgment scheme   77
          4.2.2   Flow and congestion control   78
          4.2.3   Advantages of the window-based AIMD mechanism   81
    4.3   TCP-friendly AIMD Parameters   81
          4.3.1   One TCP and one AIMD flow   82
          4.3.2   Multi-class AIMD flows   85
          4.3.3   Variable packet size and *rtt*   86
          4.3.4   Comparison with other binomial schemes   86
    4.4   Properties of AIMD   87
          4.4.1   AIMD effectiveness   87
          4.4.2   AIMD responsiveness   90
          4.4.3   Practical implications   93
          4.4.4   An enhanced AIMD algorithm – DTAIMD   94
    4.5   Case Study: Multimedia Playback Applications with
          Service Differentiation   95
          4.5.1   Multimedia playback applications   95
          4.5.2   Service differentiation 97
    4.6   Performance Evaluation   98
          4.6.1   Performance of AIMD algorithms   99
          4.6.2   QoS for multimedia playback applications   107
    4.7   Summary   110
    4.8   Problems   111

5 **Stability Property and Performance Bounds of the Internet**   **115**
   5.1   A Fluid-flow Model of the AIMD/RED System 117
   5.2   Stability and Fairness Analysis with Delay-free Marking 118
         5.2.1   Stability of the homogeneous AIMD/RED system 118
         5.2.2   Stability of the heterogeneous AIMD/RED system 120
         5.2.3   TCP-friendliness and differentiated services 123
         5.2.4   Numerical results 124
   5.3   Boundedness of the Homogeneous-flow AIMD/RED System with
         Time Delay 124
         5.3.1   Upper bound on window size 128
         5.3.2   Lower bound on window size and upper bound on queue
                 length 129
         5.3.3   Performance evaluation 132
   5.4   Summary 140
   5.5   Problems 140

6 **AIMD in Wireless Internet**   **143**
   6.1   Introduction 143
   6.2   Related Work 145
         6.2.1   TCP over wireless networks 145
         6.2.2   Using *rwnd* to enhance TCP performance 146
   6.3   System Model 147
         6.3.1   QoS indexes for delay-sensitive applications 147
         6.3.2   Wireless link throughput distribution 148
   6.4   Analytical Model for Window-controlled Flows 151
         6.4.1   Single flow, sufficient buffer 151
         6.4.2   Single flow, limited buffer 153
         6.4.3   Multiple flows 154
         6.4.4   Local retransmission delay 156
         6.4.5   Delay control for window-controlled protocol 158
         6.4.6   Further discussion 160
   6.5   Parameter Selection for AIMD 163
         6.5.1   TCP-friendliness 163
         6.5.2   *rwnd*, single AIMD flow 163
         6.5.3   *rwnd*s, multiple AIMD flows 164
         6.5.4   Parameter selection procedure 165
   6.6   Performance Evaluation 166
         6.6.1   Single AIMD flow 166
         6.6.2   Multiple AIMD flows 173
         6.6.3   AIMD vs. TCP 174
         6.6.4   AIMD vs. UDP 177
   6.7   Summary 178
   6.8   Problems 179

7   TCP-friendly Rate Control in Wireless Internet   181
    7.1   Introduction 181
    7.2   System Model 182
          7.2.1   Truncated ARQ scheme 184
          7.2.2   Wireless channel model 184
    7.3   Analytical Model for Rate-controlled Flows 184
          7.3.1   Link utilization and packet loss rate 186
          7.3.2   Delay performance 189
    7.4   Performance Evaluation 191
          7.4.1   Packet loss rate 193
          7.4.2   Link utilization 195
          7.4.3   Delay outage rate 195
          7.4.4   Effect of deterministic end-to-end delay 195
    7.5   Summary 198
    7.6   Problems 198

8   Multimedia Services in Wireless Random Access Networks   201
    8.1   Brief History of Random Access Technologies 201
    8.2   IEEE 802.11 Protocol 202
          8.2.1   DCF 204
          8.2.2   Ready-to-send/clear-to-send 204
    8.3   WLAN with Saturated Stations 205
          8.3.1   Throughput analysis 206
          8.3.2   Average frame service time 209
    8.4   WLAN with Unbalanced Traffic 209
          8.4.1   Analytical model 210
          8.4.2   Case study: voice capacity analysis 213
          8.4.3   Simulation results 220
    8.5   TFRC in the Mobile Hotspot 223
          8.5.1   System description 225
          8.5.2   TFRC throughput analysis 231
          8.5.3   Numerical results 234
    8.6   Summary 242
    8.7   Problems 242

Appendices   245

Appendix A   TCP and AQM Overview   247
    A.1   TCP Protocol 247
          A.1.1   TCP connection management 247
          A.1.2   TCP error control 247
          A.1.3   TCP flow control and congestion control 249

A.2   Active Queue Management   251

Appendix B   Datagram Congestion Control Protocol Overview   253
B.1   DCCP-2: TCP-like Congestion Control   253
B.2   DCCP-3: TFRC Congestion Control   254

References   255

Index   269

About the Series Editors

 **Xuemin (Sherman) Shen** received a B.Sc (1982) degree from Dalian Maritime University (China) and M.Sc (1987) and Ph.D degrees (1990) from Rutgers University, New Jersey (USA), all in electrical engineering. He is a University Research Chair Professor, Department of Electrical and Computer Engineering, University of Waterloo, Canada. His research focuses on mobility and resource management in interconnected wireless/wired networks, UWB wireless communications networks, wireless network security, wireless body area networks and vehicular ad hoc and sensor networks. He is a co-author of three books, and has published more than 400 papers and book chapters in wireless communications and networks, control and filtering. He serves as the Tutorial Chair for IEEE ICC'08, the Technical Program Committee Chair for IEEE Globecom'07, the General Co-Chair for Chinacom'07 and QShine'06 and the Founding Chair for IEEE Communications Society Technical Committee on P2P Communications and Networking. He also serves as a Founding Area Editor for IEEE Transactions on Wireless Communications; Editor-in-Chief for Peer-to-Peer Networking and Application; Associate Editor for IEEE Transactions on Vehicular Technology; KICS/IEEE Journal of Communications and Networks, Computer Networks; ACM/Wireless Networks; and Wireless Communications and Mobile Computing (Wiley), etc. He has also served as Guest Editor for IEEE JSAC, IEEE Wireless Communications, IEEE Communications Magazine and ACM Mobile Networks and Applications, etc. Dr Shen received the Excellent Graduate Supervision Award in 2006, and the Outstanding Performance Award in 2004 and 2008 from the University of Waterloo, the Premier's Research Excellence Award (PREA) in 2003 from the Province of Ontario, Canada, and the Distinguished Performance Award in 2002 and 2007 from the Faculty of Engineering, University of Waterloo. Dr Shen is a Fellow of IEEE, and a Distinguished Lecturer of the IEEE Communications Society. He is also a registered Professional Engineer of Ontario, Canada.

Dr Yi Pan is the Chair and a Professor in the Department of Computer Science at Georgia State University, USA. Dr Pan received his B.Eng and M.Eng degrees in computer engineering from Tsinghua University, China, in 1982 and 1984, respectively, and his Ph.D degree in computer science from the University of Pittsburgh, USA, in 1991. Dr Pan's research interests include parallel and distributed computing, optical networks, wireless networks and bioinformatics. Dr Pan has published more than 100 journal papers with over 30 papers published in various IEEE journals. In addition, he has published over 130 papers in refereed conferences (including IPDPS, ICPP, ICDCS, INFOCOM and GLOBECOM). He has also co-edited over 30 books. Dr Pan has served as an editor-in-chief or an editorial board member for 15 journals including five IEEE Transactions and has organized many international conferences and workshops. Dr Pan has delivered over ten keynote speeches at many international conferences. Dr Pan is an IEEE Distinguished Speaker (2000–2002), a Yamacraw Distinguished Speaker (2002), and a Shell Oil Colloquium Speaker (2002). He is listed in *Men of Achievement, Who's Who in America, Who's Who in American Education, Who's Who in Computational Science and Engineering*, and *Who's Who of Asian Americans*.

About the Authors

Lin Cai received M.A.Sc and Ph.D degrees (with Outstanding Achievement in Graduate Studies Award) in electrical and computer engineering from the University of Waterloo, Waterloo, Canada, in 2002 and 2005, respectively. Since July 2005, she has been an Assistant Professor in the Department of Electrical and Computer Engineering at the University of Victoria, British Columbia, Canada. Her research interests span several areas in wireless communications and networking, with a focus on network protocol and architecture design supporting emerging multimedia traffic over wireless, mobile, ad hoc, and sensor networks. She serves as an Associate Editor for IEEE Transactions on Vehicular Technology (2007–), EURASIP Journal on Wireless Communications and Networking (2006–), and International Journal of Sensor Networks (2006–).

Xuemin (Sherman) Shen received a B.Sc (1982) degree from Dalian Maritime University (China) and M.Sc (1987) and Ph.D degrees (1990) from Rutgers University, New Jersey (USA), all in electrical engineering. He is a University Research Chair Professor, Department of Electrical and Computer Engineering, University of Waterloo, Canada. His research focuses on mobility and resource management in interconnected wireless/wired networks, UWB wireless communications networks, wireless network security, wireless body area networks and vehicular ad hoc and sensor networks. He is a co-author of three books, and has published more than 400 papers and book chapters in wireless communications and networks, control and filtering. He serves as the Tutorial Chair for IEEE ICC'08, the Technical Program Committee Chair for IEEE Globecom'07, the General Co-Chair for Chinacom'07 and QShine'06 and the Founding Chair for IEEE Communications Society Technical Committee on P2P Communications and Networking. He also serves as a Founding Area Editor for IEEE Transactions on Wireless Communications; Editor-in-Chief for Peer-to-Peer Networking and Application; Associate Editor for IEEE Transactions on Vehicular Technology; KICS/IEEE Journal of Communications and Networks, Computer Networks; ACM/Wireless Networks; and Wireless Communications and Mobile Computing (Wiley), etc. He has also served as Guest Editor for IEEE JSAC, IEEE Wireless Communications, IEEE Communications Magazine and ACM Mobile Networks and Applications, etc. Dr Shen received the Excellent Graduate Supervision Award in

2006, and the Outstanding Performance Award in 2004 and 2008 from the University of Waterloo, the Premier's Research Excellence Award (PREA) in 2003 from the Province of Ontario, Canada, and the Distinguished Performance Award in 2002 and 2007 from the Faculty of Engineering, University of Waterloo. Dr Shen is a Fellow of IEEE, and a Distinguished Lecturer of the IEEE Communications Society. He is also a registered Professional Engineer of Ontario, Canada.

Jon W. Mark received a B.A.Sc degree from the University of Toronto in 1962, and M.Eng. and Ph.D degrees from McMaster University in 1968 and 1970, respectively, all in electrical engineering. From 1962 to 1970, he was an engineer and then a senior engineer at Canadian Westinghouse Co. Ltd., Hamilton, Ontario, Canada. In September 1970 he joined the Department of Electrical and Computer Engineering, University of Waterloo, Waterloo, Ontario, where he is currently a Distinguished Professor Emeritus. He served as the Department Chairman during the period July 1984–June 1990. In 1996 he established the Centre for Wireless Communications (CWC) at the University of Waterloo and is currently serving as its founding Director. Dr Mark has been on sabbatical leave at the following places: IBM Thomas J. Watson Research Center, Yorktown Heights, NY, as a Visiting Research Scientist (1976–77); AT&T Bell Laboratories, Murray Hill, NJ, as a Resident Consultant (1982–83); Laboratoire MASI, Université Pierre et Marie Curie, Paris, France, as an Invited Professor (1990–91); and Department of Electrical Engineering, National University of Singapore, as a Visiting Professor (1994–95). He has previously worked in the areas of adaptive equalization, image and video coding, spread spectrum communications, computer communication networks, ATM switch design and traffic management. His current research interests are in broadband wireless communications and networking, resource and mobility management, and cross domain interworking. He is a co-author of the text entitled *Wireless Communications and Networking* (Prentice-Hall, 2003), and the book entitled *Wireless Broadband Networks* (Wiley, 2009). A Fellow of the Canadian Academy of Engineering and a Life Fellow of IEEE, Dr Mark is the recipient of the 2000 Canadian Award for Telecommunications Research and the 2000 Award of Merit of the Education Foundation of the Federation of Chinese Canadian Professionals. He was an editor of IEEE Transactions on Communications (1983–1990), a member of the Inter-Society Steering Committee of the IEEE/ACM Transactions on Networking (1992–2003), a member of the IEEE Communications Society Awards Committee (1995–1998), an editor of Wireless Networks (1993–2004), and an associate editor of Telecommunication Systems (1994–2004). He is currently a member of the Advisory Board of the Wiley Series Advanced Texts in Communications and Networking.

Preface

With the ever-increasing demand for Internet access anywhere, anytime, the Internet and wireless systems are converging to a ubiquitous information transport infrastructure. Future proliferation of wireless Internet depends on its ability to support heterogeneous multimedia applications. Unlike traditional data applications, multimedia applications have more stringent quality of service (QoS) requirements. It is critically important to ensure the efficiency, fairness and stability of the wireless Internet, and provide satisfactory QoS for various multimedia applications. This is indeed a very challenging task.

Compared with its wired counterpart, the wireless channel exhibits much more severe impairments in the form of multipath fading, path loss, shadowing, multiple access interference, etc., which limit the usable spectrum. As a result, the wireless link is typically the bottleneck in a cross-wired/wireless-domain connection. To enhance spectral utilization, it is necessary to mitigate channel impairments using sophisticated error control and diversity techniques. Nevertheless, the signals emerging from a conditioned channel still sustain variations so that the instantaneous throughput of a wireless link is a random variable.

Emerging heterogeneous multimedia applications have different QoS requirements and traffic characteristics. There are two categories of multimedia applications: non-elastic ones with strong Service-Level Agreements (SLA), and elastic ones, which can tolerate graceful quality degradation when necessary. For non-elastic multimedia applications like Internet Protocol Television (IPTV), it is necessary to allocate dedicated network resources and deploy admission control algorithms to ensure their QoS, so the cost for supporting them is relatively high. Elastic multimedia applications can offer a degraded service when network congestion occurs. With end-to-end congestion control, they can fairly share network resources in a distributed manner. Thus, the cost for supporting elastic applications is typically low.

Traditional queueing theory works for modeling the input traffic arrival process and the output packet transmission process appropriately. For non-elastic multimedia applications over time-varying wireless links, a proper traffic model and wireless channel model can be used to quantify the queueing behavior. For elastic applications, however, the traffic arrival rate is under closed-loop congestion control in the end-systems and it is affected by the network conditions. Thus, there is no predetermined traffic model available to describe them and the traditional queueing theory cannot be effectively applied. In addition, a wireless channel is broadcast in nature, and random access protocols have been used in some wireless networks such as the IEEE 802.11 Wireless Local Area Network (WLAN). For transmissions over these wireless networks, the service rate depends on the contention levels in the network. The interaction of transmission errors, link layer contentions and end-to-end congestion control further complicates the performance modeling and analysis. Therefore, it is critical to develop a new analytical framework to quantify their performance.

The present text aims at providing analytical tools for network engineers, researchers and graduate and senior undergraduate students working in the wireless networking area. The materials in the text can be considered as a one-semester graduate course. The students who take the course would have already taken courses in the principles of data communication networks, probability and stochastic processes.

The book begins with an overview of the wireless Internet architecture, and presents the main challenges in supporting multimedia services in wireless Internet.

To understand the performance over wireless bottleneck links, first of all it is necessary to have a good knowledge of the characteristics of the wireless channels and, more importantly, how the channel profiles affect the network performance. To facilitate performance analysis and simulations, tractable and reasonably accurate channel models are essential. In Chapter 2, we introduce Markov chain-based wireless channel models that encapsulate fast fading, frequency selective fading and shadowing.

For non-elastic multimedia traffic, traffic models can be developed and used to analyze the bottleneck queueing performance. In Chapter 3, we first introduce the traffic models for popular multimedia applications such as Voice over IP (VoIP) and Moving Picture Experts Group (MPEG)-encoded video, and then an analytical framework to quantify the queue distribution for multimedia flows over wired and/or wireless links, with single-class or heterogeneous traffic. How to use the simple class-based queueing (CBQ) scheme to ensure the QoS for multiple-class traffic is also discussed.

Performance modeling and analysis for end-to-end congestion-controlled, elastic multimedia traffic over wireless links is much more challenging. We address the performance of different types of congestion control protocols in four chapters: Chapter 4 through Chapter 8.

The dominant transport layer protocol in the Internet is Transmission Control Protocol (TCP), which deploys the Additive Increase and Multiplicative Decrease (AIMD) congestion control mechanism. Since TCP was originally designed for bulk data transfer, new transport layer protocols with TCP-friendly congestion control mechanisms have been proposed to support emerging multimedia applications. There are two paradigms of TCP-friendly flow and congestion control: TCP-like window control and equation-based rate control. The representative protocols for these two paradigms are the AIMD protocol, which employs a window-based generic AIMD congestion control mechanism, and the TCP-Friendly Rate Control (TFRC) protocol, which employs an equation-based rate control mechanism.

In Chapter 4, the AIMD protocol, which inherits the congestion control mechanism of TCP and allows the adjustment of a pair of parameters to provide more flexible services for multimedia applications, is described. The necessary and sufficient TCP-friendly condition for AIMD controlled flows is derived. The effectiveness and responsiveness properties of the AIMD algorithms and the practical implications are also studied.

It has been discovered that AIMD congestion control cannot ensure network asymptotic stability. However, so long as the end-systems do not overshoot the available bandwidth too severely, the overall system efficiency can still be very high, and the packet loss rate and queueing delay can still be well bounded. Chapter 5 derives the performance bounds of TCP- and AIMD-controlled flows in the Internet. The theoretical results reveal which system parameters contribute to higher oscillations of the network and how to effectively ensure system efficiency with bounded delay.

Chapter 6 develops an analytical framework for quantifying the QoS performance of window-controlled flows (e.g., TCP and AIMD) in wireless Internet. The end-to-end delay distribution and packet loss rate of window-controlled flows over hybrid wireless and wireline networks are analytically obtained, and a delay control scheme is introduced. Simulation results are given to validate the analysis, demonstrate the feasibility of the delay control approach and show that the TCP-friendly AIMD protocol can even outperform the unresponsive User Datagram Protocol (UDP) for supporting multimedia applications in wireless Internet; this is mainly due to the self-clocking property of the window control protocols.

The TCP-Friendly Rate Control (TFRC) protocol is introduced in Chapter 7. Discrete-time Markov chains (DTMC) are developed to study the performance of TFRC-controlled flows in wireless Internet. The analysis results reveal how to ensure the delay and loss for cross-domain TFRC flows.

In Chapter 8, we further study the QoS performance in WLANs and integrated Wireless Wide Area Networks (WWANs)/WLANs. Since WLANs use a random-access medium access control (MAC) protocol, the access control mechanism and collisions observed in the link layer interact with the end-to-end congestion control, which further complicates the performance study. We first analyze the link layer performance of WLANs, considering different cases with saturated or unsaturated senders, and balanced or unbalanced traffic. We then address the more challenging research issues of how to quantify the end-to-end performance of congestion controlled flows over WLANs and integrated WWANs/WLANs.

The convenience of anytime, anywhere wireless access technologies and the escalated demand for multimedia services are the main driving force for future growth of wireless Internet. Fast evolving wireless communication technologies will enable many new multimedia applications that are beyond what the consumers can imagine today. Systematic study of network performance presented in this text is anticipated to provide insights and the right tools for wireless networking engineers and researchers, and stimulate more research activities in this important area.

We would like to thank Dr Lijun Wang, Dr Sangheon Pack, Dr Xinzhi Liu, Hong Shen, Ruonan Zhang, Fengdan Wan and Lin X. Cai for the development of the simulation programs. We would also like to thank the anonymous reviewers for their valuable comments and suggestions, which helped to improve the book. Finally, we would like to thank Sarah Tilley and Tina Ruonamma from John Wiley & Sons, Ltd for their support and help in publishing the book.

For further information please visit www.wiley.com/go/shen_multimedia

Lin Cai, Xuemin (Sherman) Shen and Jon W. Mark

1

Introduction

1.1   Convergence of Wireless Systems and the Internet

The explosive growth of the Internet in the last two decades has escalated it as the key information transport platform. Meanwhile, with the rapid advances of microelectronics and radio technologies, wireless systems are firmly increasing their popularity. The Internet Protocol (IP)-based Internet and wireless systems are converging into a ubiquitous all-IP information transport infrastructure, allowing mobile and stationary users to access the wireless Internet for multimedia services anywhere, anytime, as shown in Figure 1.1. Wireless access networks can be infrastructure-based or infrastructureless. Examples of infrastructure-based networks are wireless cellular systems/Worldwide Interoperability for Microwave Access (WiMAX), Wireless Metropolitan Area Networks (WMAN), Wireless Local Area Networks (WLAN) and Wireless Personal Area Networks (WPAN); infrastructureless ones are wireless mesh and ad hoc networks. In this text, we are particularly interested in infrastructure-based wireless cellular systems, which can provide good mobile services and currently support more than one billion customers worldwide [1]. We also address resource management and control issues in other wireless systems.

Traditionally, the IP-based Internet and wireless systems have different design principles and resource management approaches. In wireless cellular systems, since the wireless spectrum is limited and expensive,[1] to maintain the QoS of existing and handoff calls in a cell, centralized resource management and allocation schemes are used. Wireless resources are channelized, and dedicated channels are allocated to handle multimedia traffic when the connection is established, e.g., allocating certain time-slots, frequency bands, codes, or carriers in Time Division Multiple Access (TDMA), Frequency Division Multiple Access (FDMA), Code Division Multiple

---

[1]In 2001, the German government raised around $46 billion for four licenses of 2×10 MHz and two licenses of 2×5 MHz 3G spectrum.

Multimedia Services in Wireless Internet: Modeling and Analysis   Lin Cai, Xuemin Shen and Jon W. Mark
© 2009 John Wiley & Sons, Ltd

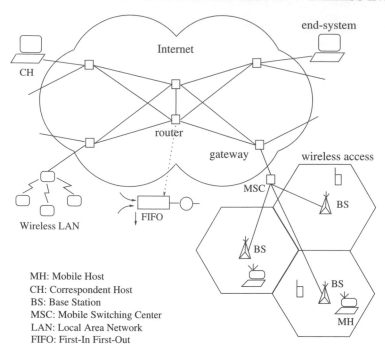

Figure 1.1 Wireless Internet.

Access (CDMA), or Orthogonal Frequency-Division Multiplexing (OFDM) systems, respectively.

Unlike wireless cellular systems, the Internet is based on the simple, robust, scalable IP protocol, which provides minimal best-effort datagram delivery services without centralized resource management and traffic control functions. As portrayed diagrammatically in Figure 1.2, when the offered traffic load is larger than the network capacity, the network power (ratio of throughput to delay) will decrease sharply and the network will be driven to deadlock and congestion collapse [2]. To efficiently and fairly share network resources in a distributed manner, the end-systems voluntarily deploy the TCP, a transport layer protocol that adjusts the sending rate according to network conditions. TCP's flow and congestion control can make the network operate near the desired operational region and maintain fairness among coexisting TCP flows. Therefore, TCP's flow and congestion control is vital for stability and integrity of the Internet.

With the rapid advances in optical and wireless communication, the Internet is becoming a more heterogeneous and disparate system: link capacity varies from several Kbps to several Gbps, with six orders of magnitude; transmission bit error rates vary from $< 10^{-9}$ to 10^{-3}, also with about six orders of magnitude; and end-to-end delay varies from several milliseconds to several seconds. An immediate question

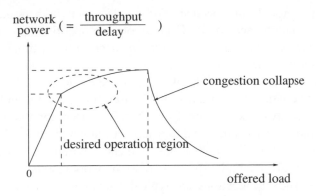

Figure 1.2 Network power.

is how to ensure the efficiency, fairness and stability of the wireless Internet, and provide satisfactory QoS for various multimedia applications, independent of the heterogeneity of communication links.

Salient background information on TCP flow and congestion control and active queue management in the Internet is summarized in Appendix A. A more comprehensive introduction can be found in [3–5].

1.2   Main Challenges in Supporting Multimedia Services

The driving force for Internet growth in the future is from emerging multimedia applications, such as audio and video streaming, IPTV and online gaming, which have a wide variety of QoS requirements. Many of these have much tighter requirements on delivery timeliness rather than object integrity. On the other hand, wireless access networks can provide convenient anytime, anywhere Internet services in a cost-effective manner (without truck-roll or rewiring costs). However, there exist many challenges to the efficient support of heterogeneous multimedia applications over wireless networks with satisfactory QoS.

First of all, a wireless channel has its inherent impairments. It suffers time-varying and location-dependent fading, shadowing, interference and ambient noise, etc., which lead to higher transmission error probabilities. Efforts from both the physical (PHY) and link layers have been conducted to mitigate the errors perceived by the upper layer protocols and applications, e.g., using coding, power control, rate control and local retransmission schemes. Although the error rate can be controlled below a certain threshold, the time to successfully transmit a packet becomes a random variable. Thus, even with a dedicated wireless channel, the service rate of a wireless link is time-varying, which introduces random delay. In addition, the wireless medium is broadcast in nature. To share the premium wireless resources, either deterministic resource allocation schemes or random access medium access control (MAC) protocols are needed. For wireless networks (e.g., IEEE 802.11 WLANs) using random access MAC protocols, competition among neighboring users that may lead to collisions can bring more uncertainty for timely transmissions. Therefore, the service rate in wireless networks is highly variable.

Different multimedia applications also have different levels of QoS requirements and traffic characteristics. For instance, IPTV users may pay for and require cable-TV-comparable quality of experience (QoE); YouTube users can tolerate much lower-quality videos and longer latency, and they pay a minimal price for the service.

For high-cost multimedia applications, dedicated wired and wireless resources can be allocated, or given a higher priority to use network resources. These high-cost (typically non-elastic) applications are typically not regulated by end-to-end flow and congestion control. Thus, the research problems are relatively simple: how to effectively reserve resource for them, and how to use admission control mechanisms to guarantee their QoS. Chapter 3 will address these issues, and Chapters 4–8 will discuss more challenging research problems for multimedia applications under flow and congestion control.

For low-cost (typically elastic) multimedia applications, the flows will share the network resources with other data traffic, so they should be regulated by end-to-end flow and congestion control to ensure network integrity and stability. However, the dominant transport layer protocol in the Internet, TCP, has been designed and engineered mainly for bulk data transfer over wireline links. It encounters the following challenges for supporting multimedia applications in wireless Internet. First, TCP provides reliable data transfer service: a TCP sender detects and retransmits corrupted or lost segments. Retransmissions by the TCP sender may introduce intolerable delay and delay jitter for delay-sensitive multimedia applications. Second, the TCP sender uses a congestion window (*cwnd*) to probe for available bandwidth[2] and respond to network congestion using an AIMD mechanism: the TCP sender additively increases its *cwnd* by one segment per round-trip time (*rtt*) if no congestion indicator is captured, and decreases the *cwnd* by half otherwise. With the increase-by-one or decrease-by-half control strategy, even an adaptive and scalable source coding scheme cannot hide the flow throughput variation, so the user-perceived audio and video quality may degrade ungracefully. Third, TCP has been designed for applications in wireline networks, where packet losses are mainly due to network congestion. Thus, packet losses seen by the TCP receiver are assumed as due to network congestion and the TCP sender is notified by congestion notifications. However, in wireless networks, packet losses may be due to transmission errors or link-layer collisions, so cutting the rate by half may underestimate the available bandwidth in wireless networks.

---

[2]Bandwidth is defined as the spectral width or capacity of a communication channel or network. Analog bandwidth is measured in Hertz (Hz) or cycles per second. Digital bandwidth is the amount or volume of data that may be sent through a channel or network, measured in bits per second. In this book, 'bandwidth' refers to digital bandwidth.

Some emerging multimedia applications use the unreliable and unresponsive UDP without flow and congestion control. However, using unresponsive transport layer protocol is not a long-term solution for high data rate and long-lived multimedia traffic: (i) it is difficult to send data using UDP in the presence of firewalls that are configured to block inbound UDP traffic; (ii) more importantly, future IP-based networks will continue to be decentralized such that network stability depends on end-to-end flow and congestion control. When unresponsive traffic occupies a large percentage of the traffic mixture, IP-based networks may be driven to congestion collapse [6]. To avoid congestion collapse, the core of the network will deploy efficient and scalable schemes to punish unresponsive traffic, e.g., dropping packets from unresponsive UDP traffic harshly when the network is under congestion [7–9]. On the other hand, although it is possible to provide congestion control at the application layer, below UDP, or by modifying other non-congestion-controlled transport layer protocols (e.g., Stream Control Transmission Protocol (SCTP) [10] and Real-time Transport Protocol (RTP) [11]), these solutions have their disadvantages and limitations: application layer congestion control has difficulties in handling Explicit Congestion Notification (ECN); congestion control below UDP needs congestion feedback from either the application or the layer below UDP, which complicates the system design; modifying SCTP or RTP may not be desirable for applications requiring minimal overhead, as discussed in [12].

Therefore, to support multimedia applications in wireless Internet, new transport layer protocols with minimal overhead, unreliable datagram delivery services, and flow and congestion control are needed. The design objectives of the new transport layer protocols for cross-domain multimedia traffic are: (i) to efficiently and fairly share highly multiplexed wired links in a distributed manner; (ii) to efficiently utilize the premium bandwidth of lightly multiplexed or dedicated wireless links; and (iii) to satisfy the QoS requirements of multimedia applications. Since TCP traffic is dominant in the Internet, the new protocol should also be TCP-friendly. *TCP-friendliness* is defined as the average throughput of non-TCP-transported flows over a large timescale does not exceed that of any conformant TCP-transported ones under the same circumstances [6].

One major difficulty for TCP-friendly congestion control is that the end-systems do not have complete knowledge of the internal network conditions, e.g., network topology, bottleneck capacity, volume of traffic sharing the bottleneck and wireless channel conditions, which are highly dynamic and difficult to anticipate. In wireless Internet, is it possible for the end-systems to obtain a fair share of the network resources and efficiently utilize them, especially the premium wireless bandwidth, with satisfactory user-perceived QoS? If so, why and how?

To meet the challenges, efforts from several layers of the protocol stack are required, and each layer tries to utilize the service from a lower layer to meet the requirements of an upper layer. Here, we use the TCP/IP reference model and focus attention on the components enclosed by the dotted ellipse shown in Figure 1.3.

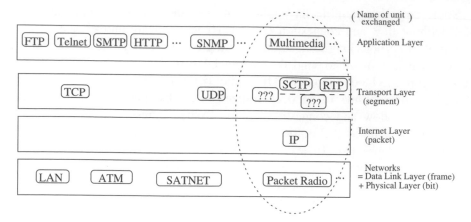

Figure 1.3 TCP/IP reference model.

Multimedia applications provide users with satisfactory visual and sound quality, based on the service provided by the transport layer protocol. The QoS provided by the transport layer protocol to the applications is measured in terms of flow throughput, packet loss rate, and end-to-end delay distribution. The transport layer protocol should probe for available bandwidth and efficiently utilize it with QoS provisioning. The link layer protocol tries to *hide* the link impairments from the upper layer protocols. The PHY layer protocol decides how information bits can be efficiently delivered through the channel.

The functionality of the relevant layers of the protocol stack can be succinctly summarized as follows. The PHY layer uses certain modulation and coding schemes to efficiently transmit information bits and ensure that the bit error rate (BER) is below a certain threshold. So long as the delay bound for delay-sensitive traffic is not violated, the link layer decides when and at what transmission power level to transmit or retransmit packets over the wireless link, so the link impairments can be as invisible to the upper layer protocols as possible. The transport layer determines when packets can be inserted into the network according to estimated network conditions. The application layer decides which information should be put in a packet to achieve the best user-perceived quality.

In the literature, scalable and error-resilient source coding schemes for multimedia applications [13–18], and link layer error correction and recovery schemes for wireless links have been extensively studied and widely deployed [19–21]. However, how the transport layer protocols appropriately support multimedia applications in ever-diversified wireless Internet and ensure their QoS needs further investigation.

1.3   Organization of the Text

Having presented an overview of the challenging issues in supporting multimedia services in wireless Internet, in Chapters 2 through 8 we provide in-depth discussions of wireless channel modeling, multimedia traffic modeling, TCP-friendly congestion control, stability and performance bounds of the Internet, QoS performance of non-congestion-controlled and congestion-controlled multimedia flows in wireless networks with dedicated resource allocation schemes and random access schemes. In the following, we briefly summarize the rationale behind the material presented in each of these chapters.

Since wireless channel characteristics and statistics determine the distribution of the service time in the wireless domain, reasonably accurate and mathematically tractable wireless channel models are essential for quantifying the performance in wireless networks. Chapter 2 introduces packet-level channel models based on Markov chains, which can be easily incorporated in the analytical frameworks and network simulation tools.

In Chapter 3, we investigate the multimedia traffic characteristics and study the traffic models for voice and video applications in the Internet. For non-congestion-controlled multimedia flows, an analytical framework is given that combines the traffic model and channel model to quantify the QoS parameters of multimedia services in wireless networks.

Chapter 4 studies the AIMD protocol, which inherits the congestion control mechanism of TCP and allows the adjustment of a pair of parameters to provide more flexible services for multimedia traffic sharing the network resources with existing TCP flows. The necessary and sufficient TCP-friendly condition for AIMD controlled flows is derived. The effectiveness and responsiveness properties of the AIMD algorithms and the practical implications are also studied.

It has been discovered that TCP and AIMD congestion control cannot ensure network asymptotic stability. However, so long as the end-systems do not overshoot the available bandwidth too severely, the overall system efficiency can still be very high, and the packet loss rate and queueing delay can still be well bounded. In this context, it is critical to investigate the theoretical bounds of the system if the network were to operate at states away from the desired equilibrium state. Chapter 5 derives an upper bound and a lower bound of the flow's congestion window size, and an upper bound of queue length. Numerical results with Matlab and simulation results with NS-2 are given to validate the correctness and demonstrate the tightness of the derived bounds. The analysis provides important insights on which system parameters contribute to higher oscillations of the system and how to effectively control system parameters to ensure system efficiency with bounded delay.

Chapter 6 develops an analytical framework for quantifying the QoS performance of window-controlled flows (e.g., TCP and AIMD) in wireless Internet. The end-to-end delay distribution and packet loss rate of window-controlled flows over hybrid wireless and wireline networks are analytically obtained, and a delay control scheme is introduced. Based on the QoS performance analysis the AIMD protocol parameters can be appropriately selected. Simulation results are given to validate the analysis, demonstrate the feasibility of the approach and show that the TCP-friendly AIMD protocol can even outperform the unresponsive UDP protocol for supporting multimedia applications in wireless Internet.

Another important TCP-friendly transport layer protocol is the TCP-Friendly Rate Control (TFRC) protocol introduced in Chapter 7. The performance of TFRC-controlled flows in wireless Internet is studied via analysis and simulation.

In Chapter 8, we further study the QoS performance in WLANs and integrated WWANs/WLANs. Since WLANs use a random access MAC protocol, the access control mechanism and collisions observed in the link layer further complicate the performance study of congestion-controlled multimedia flows. We first study the link layer performance of WLANs, and then address the more challenging research issues of how to quantify the end-to-end performance of congestion-controlled flows over WLANs and integrated WWANs/WLANs.

2

Packet-level Wireless Channel Model

2.1 Introduction

A wireless communication channel suffers inherent impairments due to multipath fading, shadowing, interference and noise. The impulse response (IR) of the wireless channel has a multipath profile that is a time-varying, environment-dependent random process. The channel IR is affected by path loss, large-scale fading and small-scale (multipath) fading.

Path loss is due to signal attenuation of the electromagnetic wave as it propagates through space. Path loss is affected by the propagation environment and medium, and it is typically represented by the path loss exponent, with a value in the range of two to six (where two is for propagation in free space, and higher values are for lossy environments).

Large-scale fading is mainly caused by random shadowing due to obstacles such as buildings and moving objects. Since the received signal is a combination of randomly delayed, reflected, scattered and diffracted components, the wireless channel is also subject to multipath fading. Small-scale fading (variations in the multipath profile of the IR) can be caused by different physical environments and movement of scatterers in the propagation surroundings. When the signal reaches the receiver by multiple paths, the signal components from different paths may be constructively or destructively superimposed. A small change in the propagation environment due to user mobility or any surrounding scatters movement will result in a significant change of the received signal level. Therefore, the channel undergoes time-varying fast fading. On the other hand, variations in the channel IR can cause the received signal-to-noise ratio (SNR) to vary differently at different frequencies, so the channel may exhibit frequency-selective fading.

Multimedia Services in Wireless Internet: Modeling and Analysis   Lin Cai, Xuemin Shen and Jon W. Mark
© 2009 John Wiley & Sons, Ltd

To fully utilize the available resource of the wireless communication channels, it is critically important to have a good knowledge of the characteristics of the wireless channels. A good packet-level channel model is not only important for the development of networking algorithms and protocols for wireless networks, but also essential for their performance analysis and optimization. Although wireless channel modeling at the physical layer has been a very active area (e.g., [22, 23]), because of complexity, it is difficult to incorporate the physical layer channel models into the design of network protocols. In this chapter, we introduce some packet-level wireless channel models that are useful for understanding protocol performance in wireless networks through performance analysis and simulation study (which is to ascertain the reasonableness of the analytical models). In the following sections, we study how to construct Markov chain-based channel models, taking into account fast fading, frequency selective fading and shadowing.

Our main goal in writing this volume is to develop analytical models to characterize the wireless propagation channel to allow for analytical performance evaluation. The reasonableness of the models is then verified by means of simulation study. The modeling and performance analysis and validation of the models via simulation carried out in this chapter sets the tone of all other chapters throughout this book.

2.2   Finite-state Markov Model for Fast Fading Channels

2.2.1   Finite-state Markov model for Rayleigh fading channels

In a typical wireless channel with multipath propagation and non-line-of-sight (NLOS) frequency-nonselective (flat) fading,[1] the received signal envelope can be modeled as Rayleigh fading. Given the modulation and coding schemes, the channel fading characteristics can be mapped to the packet level. However, with this approach, the performance analysis of the upper layer protocols becomes quite complex. Alternatively, the Rayleigh fading channel can be approximated by a finite-state Markov model, which is widely acknowledged as a reasonably accurate and mathematically tractable approach [24–27]. For notational convenience we tabulate in Table 2.1 the symbols used in this section.

A finite-state Markov model can be constructed by discretizing the received instantaneous SNR into a finite number of levels. Let g_i denote the ith state, $\mathcal{G} = \{g_1, g_2, \ldots, g_M\}$ be the set of states, and Γ_i be the ith level. The wireless channel evolves as an M-state ergodic Markov chain $G = \{G_n; n \in \mathbb{Z}_+\}$, where $G_n \in \mathcal{G} = \{g_1, g_2, \ldots, g_M\}$. Assume that all packets have the same size and that the channel remains in one state during the transmission time of a packet. The channel is in state g_i if the received SNR is between Γ_i and Γ_{i+1}. Obviously, $\Gamma_1 = 0$ and

---

[1]For a frequency-nonselective (flat) fading channel, the coherence bandwidth of the channel is larger than the bandwidth of the signal.

Table 2.1   Notation for Section 2.2.

Symbol	Description
γ	Received instantaneous signal-to-noise ratio
g_i	Channel state defined by level i and level $i + 1$
π_i	Steady-state probability of state g_i
e_i	Packet error rate of state g_i
p_e	Average packet error rate over the wireless link
$P_{i,j}$	State transition probability from state g_i to state g_j
$N(\Gamma_i)$	Level-crossing rate for SNR at Γ_i
μ	Doppler frequency shifts
λ	*Memory* of the Markov chain
t_s	Time to transmit a packet over the wireless link

$\Gamma_{M+1} = \infty$. For a Rayleigh fading channel with additive white Gaussian noise, the received instantaneous SNR, γ, is exponentially distributed:

$$p(\gamma) = \frac{1}{\gamma_0} \exp(-\gamma/\gamma_0), \tag{2.1}$$

where γ_0 is the mean of the received SNR.

The steady state probability of state g_i is given by

$$\pi_i = \int_{\Gamma_i}^{\Gamma_{i+1}} p(\gamma)d\gamma = \exp(-\Gamma_i/\gamma_0) - \exp(-\Gamma_{i+1}/\gamma_0). \tag{2.2}$$

Given a modulation scheme and a forward error correcting (FEC) code, the SNR can be mapped to BER and packet error rate (PER). Let e_i be the PER of state g_i, given by

$$e_i = \frac{1}{\pi_i} \int_{\Gamma_i}^{\Gamma_{i+1}} p(e|\gamma)p(\gamma)d\gamma, \tag{2.3}$$

where $p(e|\gamma)$ is the PER given that the SNR is equal to γ. The average PER, p_e, equals $\sum_{i=1}^{M} e_i \pi_i$.

Let $P_{i,j}$ be the state transition probability from state g_i to g_j. Let t_s be the time duration of a slot. Assume that there is no state transition within a slot, and that state transitions only occur between adjacent states, i.e., $P_{i,j} = 0$ if $|i - j| > 1$. With a slowly fading channel, assume that $N(\Gamma_{i+1})t_s$ and $N(\Gamma_i)t_s$ are less than π_i. The state transition probabilities can be approximated as [26]

$$P_{i,i+1} \approx N(\Gamma_{i+1})t_s/\pi_i, \quad i = 1, 2, \ldots, M - 1 \tag{2.4}$$

$$P_{i,i-1} \approx N(\Gamma_i)t_s/\pi_i, \quad i = 2, 3, \ldots, M, \tag{2.5}$$

where $N(\Gamma_i)$ is the level-crossing rate of the Rayleigh fading envelope at Γ_i. Level-crossing rate, which describes the frequency of channel fading, can be calculated according to the Doppler frequency shift (μ) introduced by mobility and the normalized threshold Γ_i/γ_0 [28]:

$$N(\Gamma_i) = \sqrt{2\pi}\, \mu \, \sqrt{\Gamma_i/\gamma_0} \, \exp(-\Gamma_i/\gamma_0). \tag{2.6}$$

If the Markov chain is aperiodic and indecomposable, the probability distribution of G_n will converge geometrically to the steady-state probability distribution $\pi = [\pi_1 \cdots \pi_M]^T$. The convergent speed is bounded by the second dominant eigenvalue, λ, of the channel state transition matrix [29]:

$$|\Pr\{G_n = g_i | G_1\} - \pi_i| \leq K \cdot \lambda^n, \tag{2.7}$$

where K is a constant determined by the initial state G_1. Therefore, the *memory* of a Markov chain is defined as the second dominant eigenvalue, λ, of the state transition matrix. The Markov chain's *memory* of a typical wireless channel is in the range $[0, 1)$. A channel with a positive λ means that the probability of remaining in a given state is greater than the stationary probability of that state. A channel with λ equal to zero means that the next state is independent of the current state.

Using a similar approach, finite-state Markov models can be used to approximate wireless channels with other fast fading statistics (e.g., channels with Nakagami-m distribution).

2.2.2 Mapping finite-state Markov chain to two-state Markov chain

To simplify the computational complexity for the upper layer protocol analysis, the discrete time finite-state Markov chain (FSMC) can be mapped to a two-state Markov chain with a good state (g) and a bad state (b); in the good state, the packet transmission error probability is zero, and in the bad state, the packet error probability is one. The two-state Markov chain is also referred to as the Gilbert–Elliott model [24, 25].

Each of the states in an M-state Markov chain can be viewed as good or bad with a certain probability. Consider mapping the ith state, g_i, in which the SNR falls between Γ_i and Γ_{i+1}, into a good state, (i, g), and a bad state, (i, b). As shown in Figure 2.1, an M-state Markov chain can be mapped to a two-state Markov chain by splitting each state into two substates, as indicated above, to facilitate the derivation of the state transition probabilities from the original Markov chain. The state transition probabilities from $(i, \cdot)$ to (j, g) and (j, b) are respectively $P_{i,j}(1 - e_j)$ and $P_{i,j}e_j$, and the steady state probabilities of (i, g) and (i, b) are $\pi_i(1 - e_i)$ and $\pi_i e_i$, respectively.

Then, all the states $(\cdot, g)$ are combined into a single state g, and all the states $(\cdot, b)$ are combined into a state b. The state transition probability from state g to state b equals

$$\left(\sum_{i=1}^{M} \pi_i(1 - e_i) \sum_{j=i-1}^{i+1} P_{i,j}e_j \right) \Big/ \left(\sum_{i=1}^{M} \pi_i(1 - e_i) \right),$$

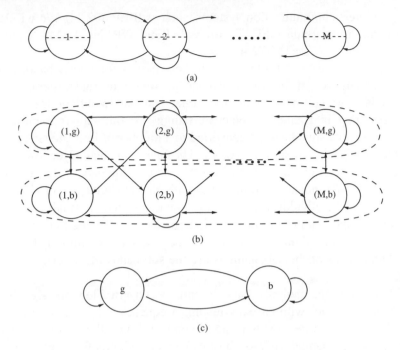

Figure 2.1 Mapping a finite-state Markov chain to a two-state Markov chain.

and the state transition probability from b to g equals

$$\left(\sum_{i=1}^{M} \pi_i e_i \sum_{j=i-1}^{i+1} P_{i,j}(1 - e_j) \right) \bigg/ \left(\sum_{i=1}^{M} \pi_i e_i \right).$$

The steady state probabilities of states g and b are equal to $1 - p_e$ and p_e, respectively. The two-state Markov chain has two eigenvalues: 1 and λ. λ is the memory of the Markov chain and it measures the burstiness of transmission errors.

The wireless fading channel can be reasonably modeled using a Markov process. For packet or multi-level transmission, the Markov process can be approximated by a finite-state Markov chain. Nevertheless, a multi-state Markov chain does not admit easy analysis of the upper layer protocols. Mapping the finite-state Markov chain to a two-state Markov chain can facilitate the upper layer protocol performance analysis.

2.3 Channel Model for Frequency-selective Fading Wireless Channels

To efficiently utilize the wireless bandwidth and effectively mitigate the intersymbol interference (ISI) associated with high data rate transmissions, multi-carrier technologies, especially OFDM, have been proposed and widely deployed in emerging wireless communication systems, such as IEEE 802.11a/g/n wireless local area networks, Multi-Band OFDM Ultrawide-Band WPANs, WiMAX and mobile broadband wireless access (MBWA) systems, etc.

In the previous section, finite-state Markov models for narrowband, flat-fading wireless channels were discussed. However, for high rate transmissions, the signal bandwidth is large compared with the coherence bandwidth of the propagation channel, resulting in frequency-selective fading. To mitigate frequency-selective fading, the high rate signal can be converted to a number of low rate signals and sent over different subchannels simultaneously. The energy of each of the low rate signals is carried by a subcarrier, each of which is the center frequency of the corresponding subchannel. The significance of transmitting a high rate signal as a group of low rate signals is that the bandwidth of each of the low rate signals is small compared with the coherence bandwidth of the propagation channel, so that the fading experienced by each of the low rate signals is confined to the frequencies within that sub-band. In fact, the bandwidths of the subchannels are the sub-bands of the original wideband channel.

OFDM is one of the methods that can send the converted low rate signals over the subchannels in parallel without experiencing frequency-selective fading. Since the bandwidth of the low rate signals is small compared with the coherence bandwidth of the propagation channel, each of the subchannels exhibits frequency-nonselective (flat) fading. Note that partitioning the channel bandwidth into sub-bands does not alter the intrinsic coherence bandwidth of the propagation channel. That is, the bandwidth of each of the low rate signals being transmitted over the subchannels is now small compared with the coherence bandwidth of the propagation channel. This means that the duration of the low rate signals is large compared with the coherence time of the propagation channel, so that the signals emerging from the subchannels will not suffer ISI. From the point of view of the converted low rate signals, the embedding of OFDM to the physical propagation channel creates an effective communications channel consisting of a group of orthogonal subchannels, each of which is supported by a subcarrier with a frequency separation greater than or equal to the width of an individual subchannel. The frequency separation chosen in this way is the reason why the parallel subchannels are orthogonal to each other. We refer to this effective communication channel as the *OFDM channel*.

We assume that each of the subchannels exhibits identical Nakagami-m slow fading. As such, adjacent subchannels are correlated. To address packet transmission, in this section we introduce a packet-level FSMC based on the premise that each of the subchannels exhibits slow Nakagami-m fading.

The remainder of this section is organized as follows. In Subsection 2.3.1, the propagation model for Nakagami-m fading signal is defined. In Subsections 2.3.2–2.3.4, we introduce the FSMC model for a multi-carrier system and how to derive the parameters of the Markov chain [30]. Simulation results are given in Subsection 2.3.5 to ascertain the reasonableness of the FSMC channel model.

2.3.1   The Nakagami-*m* fading model

Because of its good fit to the empirical fading signal measurement and generality for modeling fading with different severity, the Nakagami-*m* distribution has received much attention. According to [31], for a frequency-selective Nakagami-*m* fading channel, the magnitude of the channel frequency response is shown to be also Nakagami-*m* distributed. Therefore, each subcarrier has identical, correlated Nakagami-*m* fading.

The probability density function (PDF) of the envelope r of a Nakagami-*m* faded signal is given by

$$p(r) = \frac{2}{\Gamma(m)} \left(\frac{m}{\Omega} \right)^m r^{2m-1} e^{(-(m/\Omega)r^2)}, \tag{2.8}$$

where $\Gamma(x) = \int_0^\infty e^{-t} t^{x-1} dt$ is the Gamma function and $\Omega = E[r^2]$ is the average power of the signal. The parameter m is the fading shape factor related to the severity of the fading. For instance, for $m = 1$, the distribution becomes Rayleigh fading; for $m = \infty$, the distribution becomes an impulse, i.e., the channel becomes an additive white Gaussian noise (AWGN) channel without small-scale fading.

A propagation model for a Nakagami-*m* faded signal is developed in [32], based on the well-known Jakes' model [22]. It is shown that the envelope r can be considered as the square root of the sum of squares of m independent Rayleigh or $2m$ independent Gaussian variates so that higher order statistics, level-crossing rate (LCR) and average fade duration (AFD) can be derived.

However, the propagation model in [32] considers only narrowband flat-fading. In a wideband frequency-selective fading channel, the difference in the delay of the multiple paths gives rise to time-spread and different phase shifts on the carriers. The normalized power delay profile (PDP) can be regarded as the probability distribution of the signal delay over multiple paths. Therefore, the model proposed in [32] can be extended as follows.

Let $x_i(t)$ and $y_i(t)$ represent, respectively, the in-phase and quadrature components of the ith narrowband signal, each composed of N multipath waves:

$$x_i(t) = \sqrt{2}\sigma \sum_{j=1}^{N} a_{ij} \cos(\omega_{ij} t + \phi_{ij}), \tag{2.9}$$

$$y_i(t) = \sqrt{2}\sigma \sum_{j=1}^{N} a_{ij} \sin(\omega_{ij} t + \phi_{ij}), \tag{2.10}$$

where N is the number of sinusoidal waves, assumed to be large enough so that $x_i(t)$ and $y_i(t)$ can be approximated as Gaussian processes according to the central limit theorem. The coefficient a_{ij} satisfies the ensemble average $\langle \sum_{j=1}^{N} a_{ij}^2 \rangle = 1$, σ is the standard deviation of x_i and y_i, $\omega_{ij} = 2\pi f_m \cos(\varphi_{ij})$ is the Doppler frequency of the jth wave, $f_m = vf_c/c$, where v is the velocity of the mobile user, f_c is the carrier frequency and c is the speed of light. Hence, $\omega_{ij} t$ is the phase shift in the received wave due to the Doppler frequency. ϕ_{ij} is defined as $\phi_{ij} = 2\pi \tau_{ij} C_f f'$, where τ_{ij} is the random excessive delay of the jth path, $C_f = 100$ MHz is a constant and f' is the frequency of the carrier in units of 100 MHz. The PDP is normalized to obtain the probability distribution of τ_{ij}. For example, for the exponential decaying PDP, the probability of the excessive delay can be approximated as $p(\tau) = (1/\beta) \exp(-\tau/\beta)$, where $\beta = \tau_{\rm rms} = \sqrt{E[\tau^2] - E^2[\tau]}$ is the root mean square of the delay.

Let $r_0^2 = x_0^2$, or equivalently $r_0^2 = y_0^2$, and $r_i^2 = x_i^2 + y_i^2$. It is well known that r_0 is semipositive Gaussian distributed whereas the r_i's are Rayleigh distributed. The envelope r of the faded signal is defined as

$$r^2 = r_0^2 + \sum_{i=1}^{m-1/2} r_i^2, \quad 2m \text{ is odd,} \tag{2.11}$$

$$r^2 = \sum_{i=1}^{m} r_i^2, \quad 2m \text{ is even.} \tag{2.12}$$

As proved in [32], r fits the Nakagami-m distribution exactly, with parameters m and $\Omega = 2m\sigma^2$.

2.3.2 LCR in the frequency domain

In a frequency-selective fading channel, the coherence bandwidth B_c, which is inversely proportional to $\tau_{\rm rms}$, indicates the correlation of the frequency response, that is, the response to subcarriers of different frequencies.

To present the variation of the amplitude of different subcarriers in the frequency domain, the LCR of the magnitude of the channel frequency response should be considered. The LCR in the frequency domain can be regarded as the counterpart of the LCR in the time domain, and it represents the expected rate at which the amplitudes of the subcarriers cross a given level R in the positive direction when the frequencies of the subcarriers increase over a certain range of bandwidth.

Here we focus on the model in (2.12). The derivation for the case in (2.11) is a direct extension and the same result can be obtained. The derivative $\dot{r}$ of r with respect to frequency f' is

$$\dot{r} = \sum_{i=1}^{m} \frac{r_i}{r} \dot{r}_i. \tag{2.13}$$

As shown in [22, 32], the derivatives $\dot{r}_i$ are Gaussian distributed random variables with zero mean and standard deviation σ_i. In addition, the σ_i's are identical for $i = 1, \ldots, m$:

$$\sigma_i = \dot{\sigma} = 2\pi \sigma \tau_{\rm rms} C_f. \tag{2.14}$$

Given the individual amplitude r_i and therefore r, the distribution of $\dot{r}$ follows a Gaussian distribution with zero mean and standard deviation $\sigma_{\dot{r}} = \dot{\sigma} = 2\pi \sigma \tau_{\rm rms} C_f$.

Thus, the PDF of the frequency derivative $\dot{r}$ conditioned on r is

$$p(\dot{r}|r) = \frac{1}{\sqrt{2\pi}\dot{\sigma}} e^{(-\dot{r}^2/2\dot{\sigma}^2)}. \tag{2.15}$$

Equation (2.15) shows that $p(\dot{r}|r)$ is actually independent of r, and hence $p(\dot{r}) = p(\dot{r}|r)$ and $p(\dot{r}, r) = p(\dot{r}) \times p(r)$.

According to the definition of LCR, $N_R = \int_0^\infty \dot{r} p(\dot{r}, r = R) d\dot{r}$. The LCR in the frequency domain can be obtained as

$$N_R = p(r = R) \int_0^\infty \dot{r} p(\dot{r}) d\dot{r} = \frac{\dot{\sigma}}{\sqrt{2\pi}} p(r = R). \tag{2.16}$$

From (2.8), (2.14) and (2.16), we can get

$$N_R = 2\sqrt{\pi} \tau_{\text{rms}} C_f \frac{m^{m-1/2}}{\Gamma(m)} \rho^{2m-1} e^{-m\rho^2}, \tag{2.17}$$

where $\rho = R/r_{\text{rms}}$, and $r_{\text{rms}} = \sqrt{\Omega} = \sqrt{E[r^2]}$ is the root mean square of the envelope r.

2.3.3   FSMC in the frequency domain

Given the LCR in the frequency domain, we can construct an FSMC to model the variation of the subcarriers in the OFDM system and generate the states of the channel in terms of SNR. The channel for each subcarrier is a flat and Nakagami-m fading channel. Let E_s be the average energy per symbol and N_0 the single-sided power spectral density. Then $\gamma = r^2 E_s/N_0$ is the post-detection SNR per symbol of one subcarrier and is Gamma distributed [31, 33]. The PDF and cumulative distribution function (CDF) are, respectively, given as:

$$p_\Gamma(\gamma) = \left(\frac{m}{\bar{\gamma}}\right)^m \frac{\gamma^{(m-1)}}{\Gamma(m)} e^{-(m/\bar{\gamma})\gamma}, \tag{2.18}$$

$$F_\Gamma(\gamma) = \frac{\gamma(m, (m/\bar{\gamma})\gamma)}{\Gamma(m)}, \tag{2.19}$$

where $\bar{\gamma} = E[\gamma] = \Omega E_s/N_0$ is the average SNR.

In the transmission of high rate signals over the OFDM channel, a packet is divided into subpackets and the subpackets are transmitted over the subchannels at a fraction of the original rate. In the following, we focus on the error model of subpackets. Given the interleaving and error correction coding schemes used, the error profile of packets can be derived from the error profile of subpackets. With the idea of discrete channel state, we define K partitions for γ with thresholds $\Gamma_k, k = 0, 1, \ldots, K$ and $\Gamma_0 = 0, \Gamma_K = \infty$. Because of the slow fading in the time domain, during the period of a subpacket transmission, a subchannel stays in one state and therefore the error probability of the subpacket can be estimated by the SNR of that state [26, 27, 33, 34].

Because of the slow Nakagami-m fading, adjacent subchannels are correlated. Considering the correlation between the neighboring subchannels in the frequency domain, given the state of one subchannel, the state of the next subchannel can be estimated using the transition probability. Since the frequency difference between two adjacent subcarriers is small enough, the transition from one subchannel to the next subchannel can only happen to the current or the adjacent states (i.e., the frequency interval Δf is much smaller than the coherence bandwidth B_c). Consequently, the packet-level first-order FSMC in the frequency domain is established. The one-step transition in this model corresponds to the channel state transition from one subchannel to the next and the transition matrix P can be obtained from the LCR in the frequency domain.

In the following, the FSMC in the frequency domain is determined in detail by the state partitioning, state transition matrix and error probability vector.

SNR partitioning for Nakagami-m fading

The SNR range of each state should be large enough so that the channel stays in one state during a packet transmission but not too large, so that the PER can be estimated by the average symbol error rate of the state [27]. Iskander in [33] has provided a justification for using the equal probability method (EPM) to perform partitioning. We thus use the approach in [33] here. In a K-state system, the steady-state probabilities, π_k, $k = 0, 1, \ldots, K - 1$, of the states are all equal, i.e.,

$$\pi_k = \int_{\Gamma_k}^{\Gamma_{k+1}} p_\Gamma(\gamma)d\gamma = F_\Gamma(\Gamma_{k+1}) - F_\Gamma(\Gamma_k) = \frac{1}{K}. \tag{2.20}$$

From (2.19) and (2.20), we can obtain Γ_k by solving the following equations numerically, e.g., using the bisection method:

$$\gamma\left(m, \frac{m}{\bar{\gamma}}\Gamma_{k+1}\right) - \gamma\left(m, \frac{m}{\bar{\gamma}}\Gamma_k\right) = \frac{\Gamma(m)}{K}, \quad k = 1, 2, \ldots, K - 1, \tag{2.21}$$

with $\Gamma_0 = 0$, $\Gamma_K = \infty$.

State transition probabilities

Let $t_{i,j}$ denote the transition probability from states s_i of one subchannel to states s_j of the next subchannel. The transition probability $t_{k,k+1}$ is approximated by the ratio of the LCR at the threshold Γ_{k+1} and the average number of subchannels staying in state s_k. From (2.17), the LCR at Γ_k in the frequency domain can be obtained as

$$N(\Gamma_k) = 2\sqrt{\pi}\frac{\tau_{\text{rms}}C_f}{\Gamma(m)}\left(\frac{m}{\bar{\gamma}}\Gamma_k\right)^{m-\frac{1}{2}}e^{-(m/\bar{\gamma})\Gamma_k}. \tag{2.22}$$

Thus, the transition probabilities can be approximated as

$$t_{k,k-1} \approx \frac{N(\Gamma_k)}{L\pi_k}, \quad k = 1, 2, \ldots, K - 1, \tag{2.23}$$

$$t_{k,k+1} \approx \frac{N(\Gamma_{k+1})}{L\pi_k}, \quad k = 0, 1, 2, \ldots, K - 2, \tag{2.24}$$

where L is the total number of subchannels in the bandwidth of C_f so that $L\pi_k$ is the average number of subchannels with SNR in state s_k. The LCR defined in (2.22) is the expected number of times the SNR crossing the threshold Γ_k in the bandwidth of C_f. Thus, (2.23) gives the probability of level crossing, or equivalently the probability of state transition from s_k to s_{k-1}. Equation (2.24) is obtained in the same way. The remaining probabilities are given by

$$t_{0,0} = 1 - t_{0,1},$$

$$t_{K-1,K-1} = 1 - t_{K-1,K-2}, \tag{2.25}$$

$$t_{k,k} = 1 - t_{k,k-1} - t_{k,k+1}, \quad k = 1, 2, \ldots, K - 2.$$

Error probability vector

The average symbol error probability associated with state s_k of a subchannel is given by

$$e_k = \frac{1}{\pi_k} \int_{\Gamma_k}^{\Gamma_{k+1}} e(\gamma) p_\Gamma(\gamma) d\gamma, \tag{2.26}$$

where $e(\gamma)$ is the error probability given the SNR equal to γ, and $p_\Gamma(\gamma)$ is the Gamma distribution as in (2.18). For binary phase-shift keying (BPSK) with coherent detection, the resultant BER is shown by (8), (9) and (15) in [33].

In practice OFDM systems should have a frequency separation between subcarriers sufficiently small compared with the coherence bandwidth. Should the bandwidth of the subchannels be larger than the coherence bandwidth, virtual subcarriers may be inserted between the real subcarriers so that the difference of two neighboring virtual subcarriers is much smaller than the coherence bandwidth. Thus, the one-step transition matrix P_v between virtual subchannels is a tridiagonal matrix, and it can be determined as above. The state transition matrix of the real subcarriers is simply $(P_v)^l$, where $l - 1$ is the number of virtual subcarriers between two real subcarriers. An OFDM system should be properly designed without entailing the insertion of virtual subcarriers. Besides, the embedding of virtual subcarriers will make demodulation extremely complex.

2.3.4   FSMC model for OFDM systems

An FSMC-based model for multi-carrier communications like OFDM should include both the time domain and the frequency domain. A time-domain FSMC

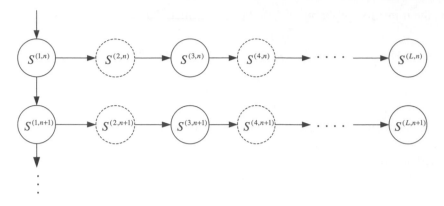

Figure 2.2 The packet-level model for Orthogonal Frequency-Division Multiplexing (OFDM) systems [30]. Reproduced by permission of ©2007 IEEE.

model [27, 33] is used to generate the states of the first subchannel at the packet transmission time, and then the FSMC in the frequency domain introduced above is used to generate the states of all the other subchannels, as shown in Figure 2.2. $S^{(l,n)}$ represents the state of the lth subchannel during the nth packet transmission. Virtual subchannels may be inserted if necessary; for example, in Figure 2.2, the even-numbered subchannels are virtual ones.

Since the channel exhibits slow fading, the system is assumed stable during the transmission of one packet to obtain the packet-level model. Meanwhile, because of the structure of the model, the correlation between the subchannels and the Nakagami-m distribution of each subchannel is maintained.

The parameters for the FSMC need to be computed once according to the channel profile and the configurations of the OFDM system. Then, in the simulation of packet transmission over the frequency-selective Nakagami-m fading channel, only $2L$ (L is the number of subchannels) uniformly distributed random variables (RVs) need to be generated for each packet. L RVs are used to determine the state transition of the subchannels, and the remaining L RVs to determine if transmission errors happen on the subchannels. This drastically reduces the computational complexity compared with the waveform bit-level simulation.

2.3.5 Simulation study

The parameters for the simulations are as follows. The velocity of the mobile user is $v = 10$ km/h and the frequency of the first subcarrier is $f_1 = 2.4$ GHz. The maximum Doppler frequency f_m is 22.22 Hz. The total bandwidth of the system is 20 MHz, which includes 512 subcarriers, and the total bit rate is 20 Mbps. Thus the interval between two subcarriers is $\Delta f = 39.062$ kHz. The packet size is assumed to be 400 bytes including the cyclic prefix (CP), resulting in a packet rate of 6250 packets/s.

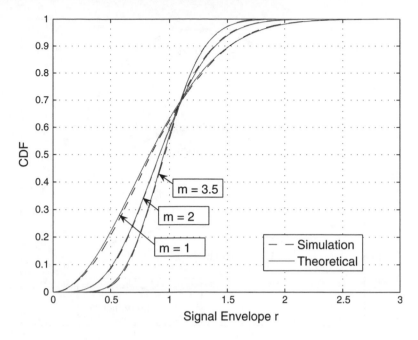

Figure 2.3 Cumulative distribution function (CDF) of the signal envelope generated by propagation model (average power $\Omega = 1$) [30]. Reproduced by permission of ©2007 IEEE.

The channel has an exponentially decaying PDP as shown in Subsection 2.3.1. The number of thresholds for discretizing the Nakagami-m distributed amplitude is 15.

Propagation model

The propagation model represented by (2.9) and (2.10) in Subsection 2.3.1 is simulated to generate the envelopes of multiple subcarriers. The number of waves is $N = 40$. The cases $m = 1, 2, 3.5$ are simulated. σ is chosen such that the Nakagami-m distribution of the generated envelope is normalized, i.e., $\Omega = 1$, so that $\sigma = 0.7071, 0.5$ and 0.3780, respectively. Each case runs for 10 seconds, corresponding to 6.25×10^4 observations.

Figure 2.3 compares the CDF between the envelope of one randomly selected subcarrier generated by the propagation model and the theoretical Nakagami-m CDF. Since a change in carrier frequency does not affect the overall fading statistics, the distribution of the subchannels should be a fixed Nakagami-m distribution, independent of the carrier frequency or Doppler shift. It can be seen that the propagation model represented by (2.9) and (2.10) generates the exact Nakagami-m distributed envelope.

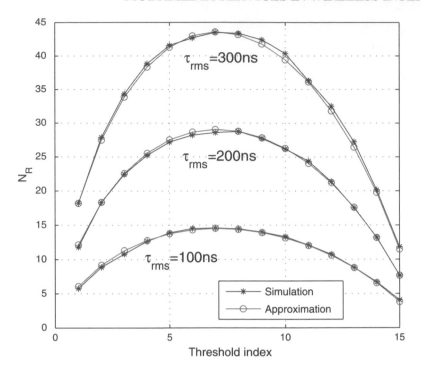

Figure 2.4 Level-crossing rate in frequency domain, $C_f = 100$ MHz [30]. Reproduced by permission of ©2007 IEEE.

LCR in frequency domain

Figure 2.4 compares the LCRs of the simulated envelope and those calculated by the theoretical expression in (2.17). The LCRs of three different channel delay spreads, τ_{rms}, (equivalently the coherence bandwidth) are simulated. The bandwidth to count the LCR is $C_f = 100$ MHz and the Nakagami-m parameter is $m = 2$. The 15 thresholds are chosen such that the Nakagami-m distributed amplitude would fit into each interval with equal probability, according to the EPM. It can be seen that the larger the delay spreads, the higher the LCR. This is expected since the channel frequency response will vary more significantly and be more frequency-selective when the coherence bandwidth decreases.

FSMC in frequency domain

We use a 16-state FSMC channel model. The amplitude of each subcarrier is generated using the propagation model and the instantaneous SNR is obtained by $\gamma = r^2 E_s/N_0$, with $E_s/N_0 = 3.01$ dB. The thresholds of the SNR intervals are chosen by the EPM such that $\pi_0 = \pi_1 = \cdots = \pi_{K-1} = 1/16$. The state transitions

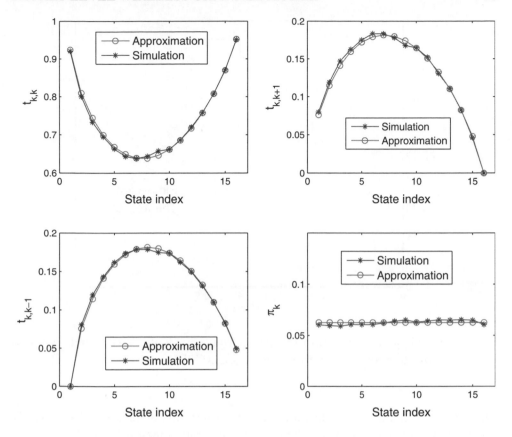

Figure 2.5 Analytical and simulation results of $t_{i,j}$ and π_k of finite-state Markov chain (FSMC) in frequency domain, $\tau_{\text{rms}} = 200$ ns, $\Delta f = 39.062$ kHz, and $m = 2$ [30]. Reproduced by permission of ©2007 IEEE.

between the SNR of neighboring subcarriers are obtained by averaging over a ten-second simulation.

The analytical approximations and simulation results for state transition probabilities, $t_{i,j}$, and steady-state probabilities, π_k, are compared in Figure 2.5, and show good agreement. The fading in the frequency domain (the frequency fading rate) is slow enough such that the assumption of state transition among neighboring states only is confirmed by the simulation results.

Finally the distribution of the states of each subcarrier generated by the FSMC model is verified. Although the state of one subchannel is generated in the frequency domain by the neighboring subchannel, they should be Nakagami-m distributed in the time domain and the SNR should have the Gamma distribution. We use the 16-state FSMC for a ten-second simulation so that a total of 6.25×10^4 observations for every subchannel is obtained, with the average SNR $\bar{\gamma} = 3.01$ dB and the fading

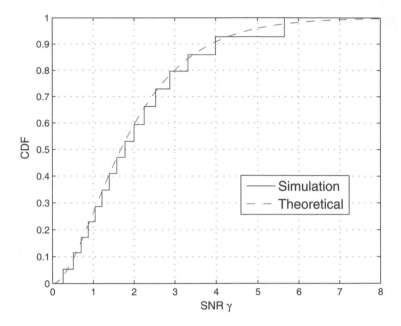

Figure 2.6   Signal-to-noise ratio (SNR) distribution of one subchannel, for $m =$ 2, $\bar{\gamma} = 2$ [30]. Reproduced by permission of ©2007 IEEE.

parameter $m = 2$. Figure 2.6 illustrates that the distribution of the SNR of a randomly chosen subchannel approximates the Gamma distribution reasonably well.

2.4   Channel Model for Indoor UWB Wireless Channels with Shadowing

So far, we have investigated how to model fast flat fading and frequency-selective fading channels, respectively. In this section, we study indoor wireless channels with a shadowing process. We focus attention on ultra-wideband (UWB) transmission technology, which has been an appealing technology for short-range, high data-rate wireless communications [35]. UWB signals occupy wide bandwidth ($\geq$ 500 MHz) at a very low transmission power ($\leq$ −41.25 dBm/MHz). Fine-resolution video, IP phone and large size files can be exchanged among UWB devices promptly. Power-constraint portable devices can also benefit from the much lower transmission power consumption of UWB.

Due to the close proximity and high density of scatterers, an indoor wireless channel contains a large number of multipath components [23] and most of the signal energy resides in the significant paths, like the line-of-sight (LOS). Efficient energy collection from these significant paths determines the range and robustness of wireless transmissions. However, in indoor environments, moving objects, such as people, may frequently penetrate the LOS and shadow off the most significant power contributions, thus considerably reducing the energy that the receiver can capture. In [36] and [37], it is shown that moving people seriously affect the performance of IEEE 802.11 2.4 GHz WLANs by introducing large variation in the received signal quality. It has also been shown [38–40] that the shadowing effect is more serious for a UWB system because of its much lower transmission power than that of other wireless systems. In addition, for UWB communications with a data rate of several hundred megabits per second, even a half second shadowing due to moving objects can affect more than 100 megabits of data, which results in severe degradation on the quality of the ongoing flows. Therefore, it is vital to consider the impact of shadowing and the resultant channel variation created by moving people on the network performance.

Since the physical layer UWB channel model and shadowing effect are too complex to be incorporated in the analytical and simulation performance evaluation of network protocols, a packet-level channel model which captures the time-variation of the UWB channel is very important and can greatly facilitate the design, analysis and simulation of the upper-layer network protocols. Networking researchers can use the model to conduct further research for UWB-based wireless networks.

The effect of shadowing on the channel multipath components can be estimated from the angular power spectral density (APSD) of the channel IR [23, 38]. The APSD refers to the magnitude of power received at a certain azimuth θ and it describes the distribution of power over the angle-of-arrival (AOA). In [38], Molisch has suggested that the angular distribution of the energy of each tap in the discrete-time channel IR be used to generate the shadowed channels and evaluate the UWB PHY systems when some propagation directions are blocked by an obstacle.[2] In [41] and [42], the realistic measurements of AOA statistics of indoor UWB channels have shown that the angular distribution of the incident rays[3] is also clustered (like the time-of-arrival characteristics). The angular clusters are distributed uniformly, and the arrivals of rays within clusters have a Laplacian distribution.

We first introduce the work reported in [43]. The APSD of UWB channel IR is derived and a modified Laplacian distribution is used to obtain an approximation of the APSD. Then the people-shadowing effect in terms of total power attenuation is evaluated analytically. Measurement results in [41, 42] and [39] confirm the analysis on APSD and shadowing effect, respectively. In addition, to study the network performance over the time-varying UWB shadowing channels, taking into consideration people walking between the transmitter (Tx) and the receiver (Rx), the approach of generating numerous random UWB channel IR realizations and performing waveform simulation is rather time-consuming. Therefore, based on the two-dimensional random walk model, a packet-level channel model can be

---

[2] This mode of operation has been recommended by the IEEE 802.15.3a standard.

[3] A ray refers to a single path in the multipath channel, and corresponds to a multipath component in the continuous-time channel IR in the IEEE 802.15.3a channel model.

established using a FSMC. The model directly generates the packet error profile for the UWB link considering channel variations. Due to the low computational requirements, the packet-level channel model can be easily incorporated into existing network simulators like NS-2 or GloMoSim, and it can also facilitate the performance/QoS analysis (link throughput, delay, etc.) of UWB networks.

The remainder of the section is organized as follows. In Subsection 2.4.1, the average APSD of the standard UWB 3a[4] channel IRs is obtained. In Subsection 2.4.2, the shadowing effect of a human body is analyzed based on the APSD and the shadowing model. The performance of multiband OFDM (MB-OFDM) UWB schemes is studied in Subsection 2.4.3. A continuous-time Markov model for the time-varying UWB channel due to people shadowing is presented in Subsection 2.4.4. Numerical results are given in Subsection 2.4.5.

2.4.1   The angular power spectral density of UWB signals

We can extend the method in [38] to investigate the APSD analytically based on the standard 3a channel model [23]. The UWB indoor channel IR defined in the 3a model is essentially a stochastic process. Each channel realization is a sample function of this stochastic process. Therefore, our objective is to obtain the ensemble average of APSD, which is defined as $P(\theta) = E[\mathbb{P}(\theta)]$, where $\mathbb{P}(\theta)$ is the APSD of a given channel realization and hence it is a random function. This average APSD gives the angular power distribution of the UWB indoor channel. Based on $P(\theta)$, the effect of people shadowing can be evaluated.

UWB 3a channel model

Using the S-V model [44], indoor UWB channel IR, composed of a series of delayed and attenuated (or faded) multipath components which are grouped into clusters and rays within each cluster, is represented by

$$h(t) = X \sum_{l=0}^{\infty} \sum_{k=1}^{\infty} a_{k,l} \delta(t - T_l - t_{k,l}), \qquad (2.27)$$

where $a_{k,l}$ is the gain of the kth ray in the lth cluster, X represents the log-normal attenuation with zero mean and variance σ_X^2, T_l is the delay of the lth cluster, and $t_{k,l}$ is the delay of the kth ray in the lth cluster relative to the cluster arrival time. The total delay of the kth ray in the lth cluster is $\tau_{k,l} = T_l + t_{k,l}$.

The cluster arrivals and the ray arrivals within each cluster are modeled as Poisson processes with arrival rates of Λ and $\lambda (\lambda > \Lambda)$, respectively. The delay of the first cluster is set as $T_0 = 0$. Because the time intervals between the cluster arrivals, $T_i - T_{i-1}$, are exponentially distributed, the cluster arrival time $T_l = \sum_{i=1}^{l}(T_i - T_{i-1})$ has a *Gamma*(l, Λ), $l = 1, 2, \ldots$ distribution. Similarly, the ray arrival time within a cluster, $t_{k,l}$, has the distribution *Gamma*(k, λ), $k = 1, 2, \ldots$.

---

[4]For notational convenience, in what follows, we will use 3a instead of IEEE 802.15.3a.

The multipath gain coefficients are modeled as independent and log-normally distributed random variables: $20 \log_{10}(|a_{k,l}|) \sim N(\mu_{k,l}, \sigma_1^2 + \sigma_2^2)$. The expected values $E[|a_{k,l}|^2]$, which are the average power delay profile of the channel, have exponential decay along the clusters (with decay factor Γ) and along the rays within the associated clusters (with factor γ):

$$\Omega_{k,l} = E[|a_{k,l}|^2] = \Omega_0 e^{-T_l/\Gamma} e^{-t_{k,l}/\gamma}, \qquad (2.28)$$

where Ω_0 is the mean energy of the first path of the first cluster. For each channel IR realization, the total energy of the multipath components is normalized such that $\sum_{l=0}^{\infty} \sum_{k=1}^{\infty} |a_{k,l}|^2 = 1$.

The constant parameters in this channel model (Λ, λ, Γ, γ, σ_1, σ_2 and σ_X) are defined by the 3a standard [23] to match the statistical characteristics of the model output with those of practical measurements. The model of CM1[5] is used throughout this work because it is the scenario with LOS existing between the transmitter Tx and the receiver Rx.

Distribution of angle-of-arrival of the rays

The AOA of each ray in the channel IR is a random variable and the distribution of the azimuthal arrival of a ray can be modeled as [38]

$$f_\theta(x|\tau) = \begin{cases} \dfrac{\tau_m}{2\pi\tau} \mathrm{rect}\left(\dfrac{\tau_m}{2\pi\tau} x\right), & 0 < \tau \le \tau_m, \ -\pi \le x < \pi \\[2mm] \dfrac{1}{2\pi}, & \tau > \tau_m, \ -\pi \le x < \pi, \end{cases} \qquad (2.29)$$

where θ is the incident angle with respect to the LOS, τ is the total delay of the ray, and $\mathrm{rect}(\cdot)$ is the rectangular function. Thus, the power received at delay τ distributes uniformly over an angular spread $\phi = \min\{2\pi\tau/\tau_m, 2\pi\}$. The parameter τ_m should be chosen such that the variance of the APSD of the channel IR is consistent with the practical measurements in [41, 42].

The measurements in [41] and [42] both indicate that the angular distribution of the rays is also clustered and the strongest cluster is almost always associated with the LOS while the other clusters are uniformly distributed over $[-\pi, \pi)$. Therefore, the ray energy distribution in (2.29) is consistent with the 3a model and the measurement. The rays with short delay should travel through short paths and arrive at the receiver within an angular range close to the LOS direction. At the same time, because of less reflection and path loss, these multipath components contain more energy, which matches the exponential decay model in the 3a standard. Particularly, the LOS component has the smallest delay and it will not distribute over any angles. On the other hand, if the rays have long delay, they usually travel through long paths and

---

[5]The IEEE 802.15.3a standard specifies four channel models. CM1 is concerned with LOS, while CM2, CM3 and CM4 are for NLOS, with different channel conditions.

more reflections. Thus, because of reflection and path loss, the range of incident arrivals should be large, with less energy and uniformly distributed over all angles.

Therefore, the early arrived rays within the first cluster in the channel IR belong to the strongest angular cluster, concentrated in the LOS direction. The other rays belong to other angular clusters. These clusters (except the LOS one) are uniformly distributed over $[-\pi, \pi)$, so that, statistically, the mean arriving angles of these clusters are immaterial. Hence, we can combine all the rays associated with these clusters together and the aggregate of their AOAs is uniformly distributed over $[-\pi, \pi)$. Thus, the angular spread of the arriving angles of the rays in the channel IR is

$$
\phi_{k,l} = \begin{cases} \dfrac{t_{k,0}}{\tau_m} 2\pi, & l = 0, \ 0 \le t_{k,0} \le \tau_m \\ 2\pi, & l = 0, \ t_{k,0} > \tau_m \\ 2\pi, & l \ge 1. \end{cases} \tag{2.30}
$$

The angular power density of the kth ray in the lth cluster can be obtained by $P_{k,l} = |a_{k,l}|^2 / \phi_{k,l}$, where $|a_{k,l}|^2$ is the energy of the ray and is a random variable with a mean value dependent on its total delay as given by (2.28). In order to evaluate the average APSD, we first consider the average angular power density of each ray conditioned on the delay $\tau_{k,l}$, which can be obtained from (2.28) and (2.30) as

$$
\overline{P_{k,l}} = E[P_{k,l}|\tau_{k,l}] = \begin{cases} E\left[\dfrac{|a_{k,0}|^2}{t_{k,0}} \dfrac{\tau_m}{2\pi} \right] = \dfrac{\tau_m \Omega_0}{2\pi} \dfrac{1}{t_{k,0}} e^{-t_{k,0}/\gamma}, & l = 0, \ 0 < t_{k,0} \le \tau_m \\[3mm] E\left[\dfrac{|a_{k,0}|^2}{2\pi} \right] = \dfrac{\Omega_0}{2\pi} e^{-t_{k,0}/\gamma}, & l = 0, \ t_{k,0} > \tau_m \\[3mm] E\left[\dfrac{|a_{k,l}|^2}{2\pi} \right] = \dfrac{\Omega_0}{2\pi} e^{-T_l/\Gamma} e^{-t_{k,l}/\gamma}, & l \ge 1, \end{cases}
$$

$$\tag{2.31}$$

which is also shown in Figure 2.7. Note that because the LOS component has no angular spread ($\phi_{1,0} = 0$), its energy should be directly added to the power density at the azimuth of $0°$.

The APSD of UWB channel IR

APSD is the composite angular power distribution, i.e., the total energy incident at an AOA θ. Since the UWB channel IR is composed of a series of delayed and attenuated rays as described in (2.27), the APSD at the azimuth θ consists of the energy contribution from all rays whose angular spread is larger than or equal to 2θ, as shown in Figure 2.7. The APSD of a channel IR realization given the cluster delay and ray delay can be represented as

$$
\mathbb{P}(\theta) = \sum_{\phi_{k,l}/2 \ge |\theta|} \overline{P_{k,l}} = \underbrace{\sum_{\phi_{k,0} \ge 2|\theta|, t_{k,0} \le \tau_m} \overline{P_{k,0}}}_{A} + \underbrace{\sum_{t_{k,0} \ge \tau_m} \overline{P_{k,0}}}_{B} + \underbrace{\sum_{l \ge 1} \overline{P_{k,l}}}_{C}, \tag{2.32}
$$

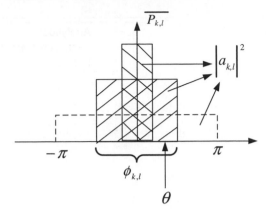

Figure 2.7 Power density of arrival angle.

where the summation A is the angular power density of the rays in the first cluster whose delay is less than τ_m but whose angular spread is larger than 2θ. Summations B and C represent the power contribution from the other rays in the first cluster and the rays in other clusters, respectively, whose angular spread covers all angles.

From (2.30), the boundaries of summation A can be transformed as $(|\theta|/\pi)\tau_m \leq t_{k,0} \leq \tau_m$. Because $t_{k,0}$ has the distribution of $Gamma(k, \lambda)$ as described in Subsection 2.4.1, we use the expected value $E[t_{k,0}] = k/\lambda$ as the approximation of $t_{k,0}$. Then the boundaries of the summation can be obtained as $k_0 \leq k \leq \lfloor \tau_m\lambda \rfloor$, where $k_0 = \max\{\lceil (\tau_m\lambda/\pi)|\theta| \rceil, 2\}$. This is because, as mentioned earlier, $k = 1$ corresponds to the LOS component and hence is excluded from the summation. $\lceil \cdot \rceil$ and $\lfloor \cdot \rfloor$ are the ceiling and floor functions, respectively, because k is an integer. Then we can evaluate the average APSD with respect to the delay terms. From (2.31) and (2.32), the average APSD (excluding the LOS component) can be obtained by (detailed derivation is referred to [43]):

$$\overline{\mathbb{P}(\theta)} = E[\mathbb{P}(\theta)] = E\left[\sum_{k_0}^{\lfloor \tau_m\lambda \rfloor} \overline{P_{k,0}} + \sum_{\lceil \tau_m\lambda \rceil}^{\infty} \overline{P_{k,0}} + \sum_{l=1}^{\infty}\sum_{k=1}^{\infty} \overline{P_{k,l}} \right]$$

$$= \underbrace{\frac{\Omega_0}{2\pi}\tau_m\lambda \sum_{k_0}^{\lfloor \tau_m\lambda \rfloor} \frac{\rho^{k-1}}{k-1}}_{\bar{A}} + \underbrace{\frac{\Omega_0}{2\pi}\frac{\rho^{\lceil \tau_m\lambda \rceil}}{1-\rho}}_{\bar{B}} + \underbrace{\frac{\Omega_0}{2\pi}(\Gamma\Lambda)(\gamma\lambda)}_{\bar{C}}, \qquad (2.33)$$

where $\rho = \lambda\gamma/(1+\lambda\gamma)$, and the terms $\bar{A}$, $\bar{B}$, $\bar{C}$ are the expectations of summations A, B, C in (2.32), respectively. In (2.33), the parameters $(\lambda, \Lambda, \gamma, \Gamma)$ are given in the 3a standard [23] or can be measured for a specific indoor environment.

The parameter Ω_0 should be chosen such that the total power contained in the multipath components is normalized to 1. Again, because $h(t)$ is a stochastic process,

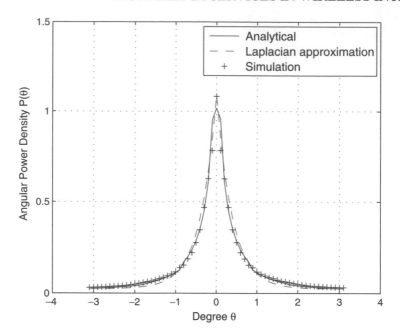

Figure 2.8 Comparison of analytical and simulation angular power spectral density (APSD) [43]. Reproduced by permission of ©2009 IEEE.

Ω_0 can be calculated by considering the average total power of the channel IR (detailed derivation is referred to [43]):

$$\Omega_0 = \frac{1}{(\gamma\lambda)(1+\Gamma\Lambda)}. \tag{2.34}$$

From (2.28), Ω_0 is the energy of the first ray in the first cluster. As mentioned earlier, the APSD at 0° should be added with the LOS component. From (2.33) and (2.34), the average APSD is obtained as

$$\overline{P(\theta)} = \begin{cases} \overline{\mathbb{P}(0)} + \dfrac{1}{(\gamma\lambda)(1+\Gamma\Lambda)}, & \theta = 0 \\[2mm] \overline{\mathbb{P}(\theta)}, & 0 < |\theta| \leq \pi. \end{cases} \tag{2.35}$$

Comparison of simulation and realistic measurements

A total of 40 channel IR realizations are generated with the 3a CM1 model, and the APSD of each IR is calculated according to [38]. $\tau_m = 14$ ns is chosen such that the standard deviation of the angular distribution is $31°$, which is the average value of the measurements in [41, 42]. The averaged APSD is shown in Figure 2.8. Using the parameters for CM1 provided in the 3a standard [23] ($\lambda = 2.5/\text{ns}$, $\Lambda =$

0.0233/ns, $\gamma = 4.3$ and $\Gamma = 7.1$), the analytical result of APSD from (2.35) is also shown in the figure. It can be seen that the analytical estimation is quite accurate.

The measurements in [41] and [42] indicate that there is an angular cluster existing in the LOS direction (the strongest one) while the other clusters distribute uniformly over $[-\pi, \pi)$. It has also been found that the relative arriving angles of the energy in one cluster are best fit to the Laplacian density $p(\theta) = \frac{1}{\sqrt{2}\sigma}e^{-(\sqrt{2}/\sigma)|\theta|}$, where, for a specific environment, the standard deviation σ can vary from 25° to 37°. Therefore, the shape of the APSD is determined by the energy distribution of the LOS angular cluster and thus should be similar to the Laplacian distribution while there is a power floor over all angles contributed from other clusters. In (2.33), the terms $\bar{B}$ and $\bar{C}$ represent the energy contributed by all angles. Based on [41,42], we use the modified Laplacian distribution as

$$P'(\theta) = \frac{1}{D}\left[\frac{1}{\sqrt{2}\sigma}\exp\left(-\frac{\sqrt{2}}{\sigma}|\theta|\right) + \bar{B} + \bar{C}\right]$$

to represent the practical measurement results. D is used to normalize $P'(\theta)$ such that $\int_{-\pi}^{\pi} P'(\theta)d\theta = 1$. Therefore, $D = 1 + 2\pi(\bar{B} + \bar{C})$ and

$$P'(\theta) = ae^{-(\sqrt{2}/\sigma)|\theta|} + b, \tag{2.36}$$

where

$$a = \frac{1}{1 + 2\pi(\bar{B} + \bar{C})}\frac{1}{\sqrt{2}\sigma} \quad \text{and} \quad b = \frac{\bar{B} + \bar{C}}{1 + 2\pi(\bar{B} + \bar{C})}.$$

The modified Laplacian distribution with $\sigma = 31°$ is also shown in Figure 2.8. The terms $\bar{B}$ and $\bar{C}$ are calculated from (2.33) with the parameters from the 3a CM1 model. Figure 2.8 illustrates that the Laplacian distribution is close to the analytical result, and fits well with the realistic measurements in [41] and [42].

2.4.2   People-shadowing effect on UWB channels

We investigate the shadowing process of the indoor UWB channel: a single scatterer, normally a person, is moving around in the area between Tx and Rx and thus blocking off the energy contribution from the significant paths. The person is modeled as a cylinder with a radius $r = 30$ cm, as an approximation of a human body, and shadows off a certain angular range, from which the transmission power cannot reach the receiver [38]. This setting of people's shadowing has been used in [39] and [38] and also recommended by the IEEE 802.15.3a group [23] to evaluate system proposals when people shadowing is considered. Based on the APSD obtained in Subsection 2.4.1, the power attenuation due to shadowing can be evaluated as follows.

Let Rx be located at the origin of the axes, and Tx on the x axis at the point $(D, 0)$, so the distance between Tx and Rx is D. The position of the moving person

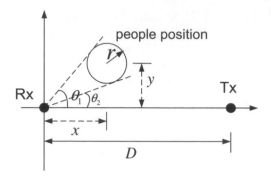

Figure 2.9 Computation of blocking angular range [43]. Reproduced by permission of ©2009 IEEE.

is denoted by (x, y), as shown in Figure 2.9. Then the angular ranges of shadowing can be obtained from geometrical computations:

$$\begin{cases} \theta_1 = \arctan\left(\dfrac{y}{x}\right) + \arcsin\left(\dfrac{r}{\sqrt{x^2 + y^2}}\right), \\[4mm] \theta_2 = \arctan\left(\dfrac{y}{x}\right) - \arcsin\left(\dfrac{r}{\sqrt{x^2 + y^2}}\right). \end{cases} \quad (2.37)$$

Finally the *remaining* power of the channel IR and thus the shadowing effect can be estimated using the APSD given in (2.35) or the modified Laplacian approximation in (2.36). If the latter is used, the remaining received power can be obtained as

$$E_r(x, y) = 1 - \int_{\theta_1}^{\theta_2} P'(\theta)d\theta$$

$$= \begin{cases} 1 - \dfrac{a\sigma}{\sqrt{2}}(e^{-(\sqrt{2}/\sigma)\theta_2} - e^{-(\sqrt{2}/\sigma)\theta_1}) - b(\theta_1 - \theta_2), & \theta_2 \geq 0 \\[4mm] 1 - \dfrac{a\sigma}{\sqrt{2}}(2 - e^{-(\sqrt{2}/\sigma)\theta_1} - e^{(\sqrt{2}/\sigma)\theta_2}) - b(\theta_1 - \theta_2), & \theta_1 \geq 0, \theta_2 < 0 \\[4mm] 1 - \dfrac{a\sigma}{\sqrt{2}}(e^{-(\sqrt{2}/\sigma)\theta_1} - e^{-(\sqrt{2}/\sigma)\theta_2}) - b(\theta_1 - \theta_2), & \theta_1 < 0, \end{cases}$$

$$(2.38)$$

where the total energy of the channel IR is normalized to 1, and θ_1 and θ_2 are calculated using (2.37). The shadowing effect (the power attenuation) in dB is

$$\chi(x, y) = 10 \log_{10}[E_r(x, y)]. \quad (2.39)$$

The numerical results of the shadowing effect when the person is penetrating the LOS along a straight path are shown in Figure 2.10. The sampling begins with the

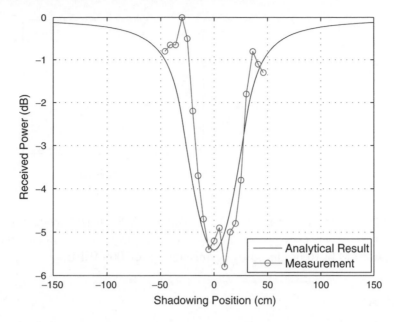

Figure 2.10 Power shadowing when people are moving along a path perpendicular to LOS [43]. Reproduced by permission of ©2009 IEEE.

person at $(0.7, 1.5)$ and ends at $(0.7, -1.5)$. Also, the realistic measurement of such a shadowing effect is given in [39], where a tank of water was used to simulate a human body and the total received energy was recorded when the obstacle was at a series of positions along the path. It can be seen that the shadowing effect obtained by the analytical model reasonably approximates the measurement. It also illustrates that the presence of moving people in the radio propagation paths introduces large attenuation on the received signal strength and thus leads to significant degradation of system performance. The shadowing effect of people randomly moving in a two-dimensional region will be given in Subsection 2.4.5.

2.4.3 Performance of MB-OFDM link with the people-shadowing effect

The important measure to evaluate UWB network performance is the achievable link throughput, which directly depends on the PER of the radio channel. Therefore, a key issue is to examine the impact of people shadowing on the PER of UWB transmission systems. The performance of an MB-OFDM system [45] is investigated in this section, and the approach can readily be extended to consider other UWB technologies like direct sequence UWB (DS-UWB).

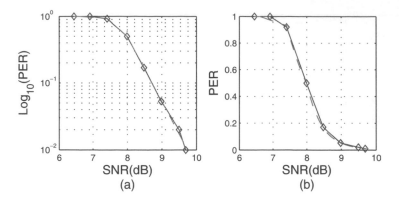

Figure 2.11 Packet error rate (PER) for Mode 1 MB-OFDM system in a CM1 channel environment ($-\diamond-$: simulation results [45]; $--$: linear regression approximation) [43]. Reproduced by permission of ©2009 IEEE.

Link budget

For MB-OFDM, when the transmission range is d, the received SNR is given by the link budget [23, 45]:

$$r(d) = P_T - L(d) - N - N_F - I, \qquad (2.40)$$

where the transmission power $P_T = -10.3$ dBm, the system noise figure $N_F = 6.6$ dB, implementation loss $I = 2.5$ dB, $N = -174 + 10 \log_{10}(R_b)$ is the thermal noise per bit where R_b is the information data rate, $L(d)$ is the path loss defined as $L(d) = 20 \log_{10}(4\pi f'_c d/c)$, where $f'_c = 3882$ MHz is the geometric center frequency of the signal.

The 90th percentile channel realization is usually of interest and is used in the simulation [45]. It corresponds to the 90th best one of 100 random channel realizations, and the resultant system performance is called 'the 90th percentile BER (or PER) performance'. Since the transmission performance is dominated by the large-scale fading and not by the signal acquisition (the multipath profile and small-scale fading) in MB-OFDM [45], the 90th percentile channel realization can be found by setting $F(X_{dB}) = Q(X_{dB}/\sigma_X) = 0.9$, where $Q(x) = \frac{1}{\sqrt{2\pi}} \int_x^\infty \exp(-t^2/2) dt$ and $\sigma_X = 3$ dB [23]. We can get $X_{dB} = -3.84$ dB.

Estimation of PER

Due to the complexity of UWB channels and transmission systems, it is very difficult to obtain accurate BER or PER in closed-form. Therefore, we will use the simulation results of PER performance provided in the MB-OFDM proposal [45] and adopt the numerical approximation method in [46] to obtain the estimation of PER from the received SNR.

Table 2.2   Coefficients and splitting points of curve fitting [43]. Reproduced by permission of ©2009 IEEE.

Segments	SNR range (dB)	a_i	b_i
1	0–3.56	−0.0378	0.1041
2	3.56–4.15	−0.4657	1.6294
3	4.15–5.84	−0.9769	3.7547

The 90th percentile PER performance [45] of 110 Mbps MB-OFDM over 3a CM1 channels is plotted in Figure 2.11. The observation suggests that we can split the PER curve into three segments. Using the approach in [46], each segment is fitted with an exponential curve (straight lines on the semi-log graph). Thus, the PER, ε, can be expressed as (with payload length of 1024 bytes)

$$\varepsilon(r) = 10^{a_i r + b_i}, \tag{2.41}$$

where r is the received SNR. The coefficients, a_i and b_i, for each segment are separately obtained using linear regression. The values of the coefficients and SNR ranges of the segments are listed in Table 2.2. As shown in Figure 2.11, the fitted curve is very close to that of the actual PER.

PER of MB-OFDM with people-shadowing effect

Considering the people-shadowing effect, the received SNR when the distance between Tx and Rx is D and the person is standing at the point (x, y) can be obtained by

$$r(x, y) = r_0 + \chi(x, y), \tag{2.42}$$

where $r_0 = r(D)$ represents the received SNR without people shadowing from (2.40) and $\chi(x, y)$ is the power attenuation by the shadowing determined by (2.38) and (2.39) (in dB and negative values). Then the PER of a MB-OFDM system with such a shadowing effect can be obtained using (2.41), with the received SNR equal to $r(x, y)$.

From the link budget analysis, the following conclusions can be drawn. First, if Tx and Rx are close enough and thus the UWB system has sufficient link margin, additional attenuation on the received power introduced by people shadowing can be tolerable and the system can still provide reliable communication. However, if the distance between Tx and Rx is large and there is insufficient link margin, the system performance will degrade significantly during people shadowing. This section focuses on the second scenario. The numerical results of SNR with people moving in the vicinity of LOS are given in Subsection 2.4.5.

2.4.4   Markov model for UWB channel with people shadowing

To describe the time-variation of an indoor UWB channel with a random shadowing process, a packet-level channel model is introduced in this section. We consider the scenario in which a person randomly enters the LOS vicinity and moves in the two-dimensional space between Tx and Rx. Because the channel states correspond to the spatial zones and the person can only walk into adjacent zones from the current one, the shadowing process can be modeled as a first-order Markov process. Therefore, similar to the FSMC channel model widely used in modeling the Rayleigh fading channel [26, 27], we can construct a Markov chain to describe the UWB shadowing channel at the packet level. The Markov channel model is characterized by two parameters: the channel states and the state transition rate for continuous-time FSMC or transition probabilities for discrete-time FSMC.

Definition of channel states

The shadowing of a randomly moving person imposes a time-varying attenuation on the total received power, which results in large-scale fading of the received SNR, as shown in (2.42). Following the approach in [27, 33], we can partition the received SNR into several non-overlapping intervals, each of which corresponds to a state of the channel.

We can draw the contour lines where each contour line corresponds to a series of people positions, (x, y), resulting in the same received SNR, r, after shadowing. This is shown later in Figure 2.13. The SNR values of the N contour lines are denoted as R_n, $n = 1, 2, \ldots, N$ and $R_{n+1} < R_n$. These contour lines divide the whole region into $N + 1$ zones. The zone between the two boundary contour lines of R_n and R_{n+1} corresponds to the SNR interval of $[R_{n+1}, R_n)$. If a person stands inside this zone, the received SNR will be $r(x, y) \in [R_{n+1}, R_n)$. This zone defines the nth channel state, denoted as S_n ($n = 1, 2, \ldots, N - 1$). State S_N corresponds to the most central zone enclosed by the Nth contour line and has the most severe shadowing effect, with the SNR interval of $[R_{N+1}, R_N)$ where R_{N+1} is the minimum received SNR. In addition, we define S_0 as the state corresponding to the zone outside the most exterior contour line R_0 and the SNR interval of $[R_1, R_0)$ where $R_0 = r(D)$ is the SNR without shadowing, which is determined by the link budget as in (2.40). State S_0 represents the condition of no shadowing, so the UWB link has the highest SNR.

Finally, following the approach in Subsection 2.2.1, the average PER of channel state S_n can be obtained as

$$\overline{\varepsilon_n} = \frac{1}{R_n - R_{n+1}} \int_{R_{n+1}}^{R_n} \varepsilon(r)dr, \tag{2.43}$$

where the PER, $\varepsilon(r)$, is given by (2.41). Furthermore, the link throughput when there is no packet error has been provided in the MB-OFDM proposal [45], denoted as H

(Mbps). Then the average throughput of state S_n can be approximated as

$$\bar{H}_n = (1 - \bar{\varepsilon}_n)H. \tag{2.44}$$

State transition rate

The transition rate between the states is determined by the area of the zones and people's mobility. In this subsection, we use a simple mobility model to build the FSMC channel model. The approach can be extended to consider other mobility models, e.g., those in [47–49].

First, the contour line of R_1 is the boundary of the shadowing region. A person arriving (entering the boundary) means that a shadowing occurs, which corresponds to the state transition from S_0 to S_1. We assume that the arrival of a person is a Poisson process with the arrival rate of λ_0, which is determined by the density and activity of the people inside the home or office. For example, in a more crowded area, we can set a higher value of λ_0, and vice versa. When the person moves out of the boundary, this person (or another person) may re-enter the region from any point later.

Second, the instantaneous speed and direction of the people vary randomly and hence the duration of the people staying in one zone is a random variable. We assume that the time the person stays inside a zone is exponentially distributed and the average duration is proportional to the area of the zone. The average duration of state S_n is $\bar{t}_n = A_n T$, where A_n is the area of the zone and T is the average duration for which the person stays in a unit area. T reflects the average movement speed of the person. Then the departure rate from state S_n is $v_n = 1/\bar{t}_n = 1/(A_n T)$. Further, we assign the transition rate to the adjacent inner zone (from S_n to S_{n+1}) to be $\lambda_n = \alpha v_n$ and the rate to the adjacent exterior zone (from S_n to S_{n-1}) to be $\mu_n = (1 - \alpha)v_n$, where $0 < \alpha < 1$ depends on the probability of a movement to the inner zone. Note that for state S_N, we have $\mu_N = v_n$. The transition rates are summarized as follows:

$$\begin{cases} \lambda_n = \lambda_0, & n = 0 \\ \lambda_n = \alpha v_n = \dfrac{\alpha}{A_n T}, & n = 1, 2, \ldots, N \\ \mu_n = (1 - \alpha)v_n = \dfrac{1 - \alpha}{A_n T}, & n = 1, 2, \ldots, N - 1 \\ \mu_n = v_n = \dfrac{1}{A_n T}, & n = N. \end{cases} \tag{2.45}$$

Given the states and the state transition rate, a continuous-time Markov model of the UWB channel can be built, as shown in Figure 2.12.

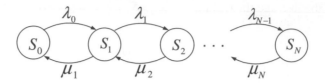

Figure 2.12   Finite-state Markov chain model for ultra-wideband channel with shadowing process.

Steady state probability

According to the equilibrium equations of the continuous-time FSMC, the steady state probability of S_n is

$$p_n = \frac{\lambda_0 \lambda_1 \cdots \lambda_{n-1}}{\mu_1 \mu_2 \cdots \mu_n} p_0 = \frac{\lambda_0}{\mu_n} \left(\frac{\alpha}{1-\alpha} \right)^{n-1} p_0, \quad n = 1, 2, \ldots, N,$$

where the λ_n and μ_n are given in (2.45) and p_0 is the steady state probability of S_0. Furthermore, based on the condition $\sum_{n=0}^{N} p_n = 1$, p_n can be derived as

$$p_n = \begin{cases} \left[1 + \lambda_0 \sum_{i=1}^{N} \frac{1}{\mu_i} \left(\frac{\alpha}{1-\alpha} \right)^{i-1} \right]^{-1}, & n = 0 \\[2em] \left[1 + \lambda_0 \sum_{i=1}^{N} \frac{1}{\mu_i} \left(\frac{\alpha}{1-\alpha} \right)^{i-1} \right]^{-1} \frac{\lambda_0}{\mu_n} \left(\frac{\alpha}{1-\alpha} \right)^{n-1}, & n = 1, 2, \ldots, N. \end{cases}$$

$$(2.46)$$

2.4.5   Numerical results

We evaluate the scenario that stationary Tx and Rx are deployed with distance $D = 10$ m. The data rate of the MB-OFDM system is 110 Mbps (according to the data rate versus range requirement of UWB systems). The payload size of a frame is 1024 bytes. The received SNR of the 90th percentile channel realization is $r_0 = 6.12$ dB from (2.40) and it is the SNR when there is no people shadowing. The Markov model of $N = 6$ states is used here.

Channel states

The received SNR with shadowing effect is calculated by (2.42) as a function of the positions of the people, and the contour lines of the received SNR are plotted in Figure 2.13. These five contour lines divide the region into six zones, which correspond to the six channel states described in Subsection 2.4.4. The SNR intervals for state S_0 to state S_5 are [5.5, 6.12), [5, 5.5), [4.5, 5), [4, 4.5), [3.5, 4) and [0, 3.5) dB, respectively. The average PER of each state is obtained by (2.43) and

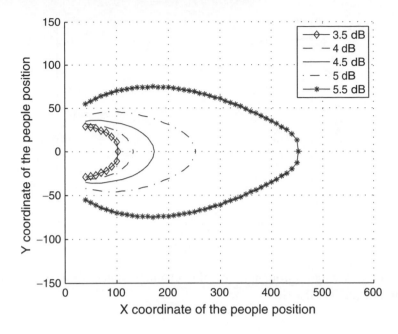

Figure 2.13 Received signal-to-noise ratio with shadowing [43]. Reproduced by permission of ©2009 IEEE.

Table 2.3 Parameters for the Markov model [43]. Reproduced by permission of ©2009 IEEE.

	n					
	0	1	2	3	4	5
Area A_n (m^2)	—	3.30	0.808	0.328	0.171	0.323
λ_n (/s)	0.500	0.067	0.275	0.678	1.300	—
μ_n (/s)	—	0.135	0.550	1.355	2.599	1.376
p_n	0.189	0.700	0.086	0.017	0.005	0.004
PER $\bar{\varepsilon}_n$	0.014	0.045	0.137	0.406	0.771	0.994

shown in Table 2.3. It can be seen that the PER increases drastically when the person blocks the LOS gradually and thus the link performance degrades severely.

If a single frame transmission is used and there is no packet error, the throughput for frames with a 1024-byte payload is $H = 83.48$ Mbps [45]. According to (2.44), the average throughput of each channel state can be obtained, as shown in Figure 2.14.

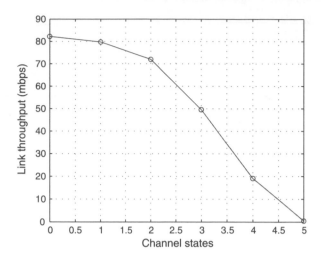

Figure 2.14   Average throughput of the channel states [43]. Reproduced by permission of ©2009 IEEE.

Transition rate

As an example of the people mobility model, assume that the people arrival rate is $\lambda_0 = 0.5$ per second (Poisson arrival) and the average duration of stay in one unit area is $T = 1$ s/m^2. The transition rate and the steady state probabilities of the Markov channel model can be obtained by (2.45) and (2.46), respectively. The results of the model parameters are listed in Table 2.3.

It can be seen that the steady state probabilities depend on the people arrival rate λ_0. When λ_0 is small, the LOS shadowing happens occasionally and, therefore, the probability of the good channel state is high; when λ_0 is large, the LOS shadowing happens frequently and the probabilities of the shadowing states increase.

Figure 2.15 illustrates the fluctuation of the throughput of the peer-to-peer MB-OFDM system. The Markov model obtained above is used to generate the random link throughput. Frequent variations of the throughput from 80 Mbps to less than 20 Mbps can be observed.

2.5   Summary

For performance analysis in wireless networks, it is of critical importance to investigate the wireless channel profile and, more importantly, its impact on upper layer protocols. For packet-switching networks, the characteristics of a wireless channel that are most relevant to network performance are packet transmission time (or rate), packet error probability, and channel correlations in the time and frequency domains.

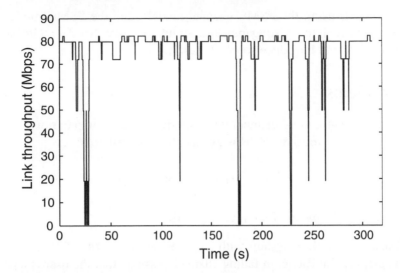

Figure 2.15 Throughput fluctuation of a Multiband Orthogonal Frequency-Division Multiplexing link due to the shadowing effect [43]. Reproduced by permission of ©2009 IEEE.

For a narrowband fading channel, because of the superposition of the multipath components and the difference in their phases, the IR is time-variant, caused by constructive and destructive interference of the multipath components. For small-scale fading that can be modeled as Rayleigh or Nakagami-m, packet-level FSMC channel models have been established to track the variation of the channel statistically [26, 27, 50].

Because the structures and the characteristics of the IR of narrowband and wideband wireless channels are different, narrowband FSMC channel models are not directly applicable for describing the small-scale fading of wideband channels. To capture the frequency-domain channel correlation, a FSMC in the frequency domain has been established in [30]. How to construct a two-dimensional FSMC to capture fast, frequency-selective fading characteristics of wideband channels is still an open issue worth further investigation.

For indoor UWB wireless networks with low mobility, without considering the shadowing effect, the channel can be considered to exhibit block-fading, i.e., it is time-invariant during the transmission of each packet but the states of different packets are independent. However, in the presence of people shadowing, the indoor UWB channels also exhibit time variation and correlation due to shadowing. A FSMC channel model has been proposed in [40, 43] to describe the impact of the shadowing effect on packet-level performance.

In summary, the FSMC channel models are very important for network analysis and simulation. Many research works have been conducted to analyze and improve network protocols using the FSMC channel models, and the FSMC models can be easily incorporated into existing network simulators like NS-2 or GloMoSim. In Chapters 6–8, we will introduce some of these works.

2.6 Problems

1. For a Rayleigh fading channel with additive white Gaussian noise, verify that the PDF of the received SNR is exponentially distributed:

$$p(\gamma) = \frac{1}{\gamma_0} \exp(-\gamma/\gamma_0), \tag{2.47}$$

where γ_0 is the mean of the received SNR.

2. Consider a wireless system with a carrier frequency of 2.54 GHz, where the channel is a flat Rayleigh fading channel. Assume that the user is a pedestrian at a speed of 1 m/s. Let γ_0 be the mean of the received SNR.

   (a) Find the LCR at γ_0.

   (b) Find the LCR at $\gamma_0/2$.

   (c) At which received SNR level is the LCR maximized?

   (d) What are the percentages of time when the received SNR is below $\gamma_0/2$ and above γ_0, respectively?

3. Consider a two-state Markov chain with zero memory. The steady state probabilities of the two states are s_1 and s_2, respectively. Find the state transition probability matrix of the Markov chain.

4. We can construct a three-state Markov chain model for a Rayleigh fading wireless channel: the channel is in state g_1, g_2 and g_3 if the received SNR is in $(0, \Gamma_1]$, $(\Gamma_1, \Gamma_2]$, and (Γ_2, ∞), respectively. For a Rayleigh fading channel with additive white Gaussian noise, the received instantaneous SNR, γ, is exponentially distributed:

$$p(\gamma) = \frac{1}{\gamma_0} \exp(-\gamma/\gamma_0),$$

where γ_0 is the mean of the received SNR. Let $N(\Gamma_1)$ and $N(\Gamma_2)$ be the LCR of the Rayleigh fading envelope at Γ_1 and Γ_2, respectively. The transmission time of a packet over the wireless channel is t_s.

   (a) Derive the steady state probabilities of the Markov chain.

   (b) Find the state transition probability matrix of the Markov chain.

   (c) Find the memory of the Markov chain.

(d) The packet error rate is e_i when the channel is in state g_i.

   (i) Find the average packet error rate.

  (ii) Find the channel throughput (i.e., given that the channel is always busy transmitting, find the number of packets being successfully transmitted over the channel per second).

(e) Convert the three-state Markov chain to a two-state Markov chain according to the method in Subsection 2.2.2.

   (i) Find the steady state probabilities of the two-state Markov chain.

  (ii) Find the state transition probability matrix of the two-state Markov chain.

5. For a frequency non-selective (flat) Nakagami-m fading channel, the PDF of the envelope r of faded signal is given by

$$p(r) = \frac{2}{\Gamma(m)} \left(\frac{m}{\Omega}\right)^m r^{2m-1} e^{(-(m/\Omega)r^2)},$$

where $\Gamma(x) = \int_0^\infty e^{-t} t^{x-1} dt$ is the Gamma function and $\Omega = E[r^2]$ is the average power of the signal.

(a) Find the LCR $N(y)$ at $y = \Omega$.

(b) To construct a three-state Markov chain with equal steady state probability for each state, how should the SNR be partitioned?

6. For a Nakagami-m channel, derive the LCR in the frequency domain.

7. Doppler shift of the channel affects the LCR in the time domain. Does it affect the LCR in the frequency domain? Describe what kind of channel has higher LCR in the frequency domain.

8. Given the multipath coefficients in (2.28) and the angular spread of the arriving angles of the rays in (2.30), verify the average angular power density of each ray expressed as below:

$$\overline{P_{k,l}} = E[P_{k,l}|\tau_{k,l}]$$

$$= \begin{cases} E\left[\dfrac{|a_{k,0}|^2}{t_{k,0}} \dfrac{\tau_m}{2\pi}\right] = \dfrac{\tau_m \Omega_0}{2\pi} \dfrac{1}{t_{k,0}} e^{-t_{k,0}/\gamma}, & l = 0, \ 0 < t_{k,0} \le \tau_m \\[3mm] E\left[\dfrac{|a_{k,0}|^2}{2\pi}\right] = \dfrac{\Omega_0}{2\pi} e^{-t_{k,0}/\gamma}, & l = 0, \ t_{k,0} > \tau_m \\[3mm] E\left[\dfrac{|a_{k,l}|^2}{2\pi}\right] = \dfrac{\Omega_0}{2\pi} e^{-T_l/\Gamma} e^{-t_{k,l}/\gamma}, & l \ge 1. \end{cases}$$

As shown in Figure (a), the distance D between the transmitter and the receiver is 10 m. A water tank (assumed to be sufficiently tall) has a rectangular shape

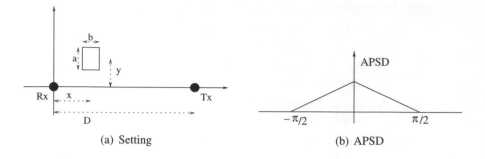

(a) Setting   (b) APSD

cross section of size $a = 30$ cm and $b = 20$ cm. The coordinates of the center of the tank are $x = 1.5$ m and $y = 1$ m. Assume that the APSD is distributed as in Figure (b). Obtain the total power being blocked by the water tank.

3

Multimedia Traffic Model

To understand the performance of multimedia traffic transmitted over heterogeneous wireless and wired channels,[1] both channel and traffic characteristics play important roles. From a queueing theory point of view, the channel characteristics determine the output service statistics, while the traffic characteristics determine the input arrival statistics. In Chapter 2, we have discussed how to model time-varying wireless channels. In this chapter, we introduce multimedia traffic models, and jointly consider the traffic model and the channel model to quantify the QoS parameters of multimedia services over a wireless channel.

The chapter is organized as follows. Section 3.1 introduces VoIP technology and the traffic model for VoIP calls. Section 3.2 presents two models for video traffic: (a) a mini-source model and (b) a two-level Markovian model. How to utilize the traffic models to quantify the queueing performance of video over wired and wireless links is addressed in Section 3.3. Section 3.4 briefly introduces scalable source coding, followed by concluding remarks in Section 3.5.

3.1 Modeling VoIP Traffic

3.1.1 VoIP

VoIP, also known as IP telephony and Internet telephony, is a set of protocols to transport voice traffic over IP-based packet-switched networks with acceptable quality of service (QoS) and reasonable cost. Efforts in transmitting voice over packet-switched networks can be traced back to the early 1970s. Since the mid 1990s, because of its cost effectiveness, IP telephony service has advanced rapidly, and is anticipated as a viable alternative to the traditional voice service over public switched telephone networks (PSTN). In addition, VoIP can efficiently provide compelling features and services, such as voice mail, voice conferencing, etc., by allowing the integrated transmission of voice and data over the same network.

[1] The terms channel and link will be used interchangeably as appropriate throughout the chapter.

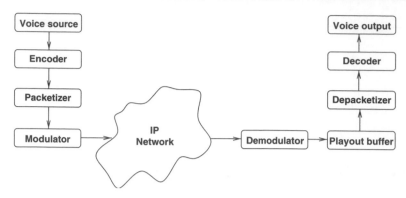

Figure 3.1 Voice over Internet Protocol system.

Analog sources such as voice and images are continuous in time and amplitude; they are highly redundant and contain unnecessary, or irrelevant, information. Efficient use of the available channel bandwidth requires the removal of redundancy and suppression of irrelevancy. Information is measured in terms of entropy. The entropy of an analog signal is infinite while that of a digital signal is finite. Since the ear and the eye can discern only a finite amount of information, it is beneficial to convert the analog signal to a digital one prior to redundancy removal.

Conversion of an analog signal to a digital format (A-to-D conversion) involves two processes: sampling and quantization. Quantization throws away unnecessary information so that a digital symbol can be represented by a finite number of signal levels, or a finite number of bits. The digital source is then encoded to remove the redundant data. The signal encoded at the transmitter has to be decoded at the receiver prior to presentation to the destination. The combination of an encoder and a decoder is referred to as a *codec*. Commercially available codecs may include other functionalities, including A-to-D conversion.

There are three indispensable VoIP components at the end-systems: codec, packetizer/depacketizer and playout buffer, as shown in Figure 3.1. The analog voice signals are digitized and then encoded into digital voice streams by the codec. Voice codecs are standardized by the International Telecommunication Union – Telecommunication (ITU–T), and include G.711 with speech rate of 64 Kbps, G.729 with 8 Kbps, G.723.1a with 5.3/6.3 Kbps, etc. Another popular voice codec is internet Low Bit Rate Codec (iLBC), which handles packet losses through graceful speech quality degradation and thus is more robust against packet losses. iLBC has been used in popular VoIP applications such as Skype.

The main attributes of some commonly used encoding schemes are listed in Table 3.1. The encoded voice stream is packetized to generate constant bit rate (CBR) audio packets, each with an appended RTP(RTCP)/UDP/IP header, where RTP and RTCP [51] are respectively the real-time transport and real-time transport control

Table 3.1   Attributes of commonly used encoding schemes.

Codec	Bit rate (Kbps)	Sample period (ms)	Payload (bytes)	Packet/s	MOS**
G.711	64	20	160	50	4.1
G.723.1a	5.3	30	20	33	3.65
G.723.1a	6.3	30	24	33	3.9
G.729	8	10	10	50*	3.7
iLBC (1)	13.33	30	50	33	***
iLBC (2)	15.20	20	38	50	4.14

*For G.729, two frames are combined into one packet; **MOS: Mean Opinion Score; ***iLBC has better quality than G.729A, and it is also more robust against packet losses.

protocols, designed to support real-time multimedia applications with stringent delay constraints and transmitted using the unreliable UDP. In addition, call setup signaling protocols, such as Session Initiation Protocol (SIP) and H.323, are used for establishing VoIP connections. SIP, defined in RFC 2543 of the Internet Engineering Task Force (IETF), is designed for Internet applications, while H.323, standardized by the ITU–T, is developed for VoIP and multimedia conferencing applications. The VoIP protocol stack is illustrated in Figure 3.2. Voice packets are transmitted over the IP network, and the reverse processes of depacketizing and decoding are accomplished at the receiver. A playout buffer is used by the receiver to smooth the speech by eliminating delay variations. Packets arriving later than the playout time will simply be discarded. Some other components such as voice/silence detector, loss/error concealment and echo canceler are also included in the system to enhance the functionality and performance of VoIP systems. The major metric to evaluate the user-perceived voice quality is Mean Opinion Score (MOS),[2] which is rated on a scale of 1 to 5.

3.1.2   VoIP traffic model

The characteristics of voice traffic largely depend on the voice codec used. Some codecs (e.g., G.711) do not implement a voice/silence detector and the resultant voice traffic is simply a CBR packet stream. For codecs with voice/silence detection, since human speech consists of sequences of alternating active intervals, or talkspurts, and silent (inactive) intervals [52], a two-state birth–death model (also called an 'on/off' model) has been widely used to describe such a packet voice source. As shown in Figure 3.3 (a), during a talkspurt ('on' state), the source generates CBR audio packets with data rate A bps; during a silent interval ('off' state), no audio packets

---

[2]To determine MOS, a number of listeners rate the quality of test sentences transmitted over the communication system as follows: (1) bad; (2) poor; (3) fair; (4) good; (5) excellent. The MOS is the arithmetic mean of all the individual scores.

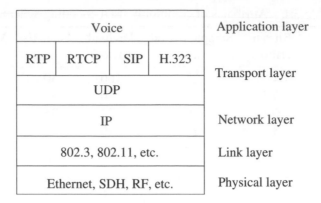

Voice				Application layer
RTP	RTCP	SIP	H.323	Transport layer
UDP				
IP				Network layer
802.3, 802.11, etc.				Link layer
Ethernet, SDH, RF, etc.				Physical layer

Figure 3.2 VoIP protocol stack.

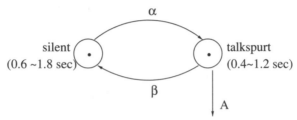

α — the rate of transition from the silent state to the talkspurt state
β — the rate of transition from the talkspurt state to the silent state
A — the constant data rate generated by the voice source during talkspurt

(a) A single voice source

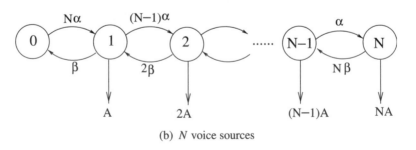

(b) *N* voice sources

Figure 3.3 Birth–death model.

are generated. The transition rate between the 'on' and 'off' states equals the inverse of the average durations of talkspurt and silent intervals.

Using the 'on/off' model, when *N* voice sources are multiplexed and sharing a channel, the total traffic can be modeled as an $(N + 1)$-state birth–death model, as

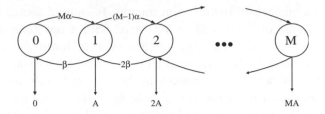

Figure 3.4 The mini-source traffic model for video.

shown in Figure 3.3 (b). The state represents the number of sources active, and the total voice traffic generated in state i equals iA bps.

Although the 'on/off' model has been widely used for packet voice traffic in the literature, it is less suitable for VoIP traffic modeling purposes, considering the popular voice codecs used in VoIP systems. For instance, in G.729 [53], during a silent period, Silence Insertion Descriptor (SID) packets are generated and transmitted to update the background noise parameters. This is because the listener may assume that transmission for the ongoing conversation has been stopped if there is no background noise during the silent period. The payloads of SID packets are typically much smaller than those of the normal voice packets during the talkspurt; however, considering the overheads of the protocol stack, the overall packet size of SID is only slightly smaller than that of active speech packets.

Therefore, for VoIP traffic, we can either be conservative and ignore the silent period and use CBR traffic to model it, or we can modify the 'on/off' model by considering the SID traffic during the silent period ('off' state). In the latter case, the traffic volume of state i equals $iA + (N - i)S$, where S represents the data rate of SID packets during the silent period.

3.2   Modeling Video Traffic

The statistics of video traffic are determined by several factors, e.g., coding schemes, video content, encoding parameters, etc. Extensive research on video traffic modeling has been conducted and a variety of models have been proposed in the literature [54–60]. In this book, we focus on the impact of traffic characteristics that are most relevant to the network performance, and pay less attention to the exact statistics of each video stream. Therefore, in this section, we introduce two relatively simple but effective traffic models that are useful for understanding video performance when it traverses a network with wired and wireless links.

3.2.1   Mini-source video model

A classic video model for the video source rate is the continuous time Markov model based on mini sources [54]. A variable bit rate video source was modeled as a superposition of a number of 'on/off' mini sources. Each mini source independently alternates between an 'off' state and an 'on' state, with A bps generated during the 'on' state. A mini source does not generate any packets during the 'off' state. The average residence times in the 'off' state and 'on' state are $1/\alpha$ and $1/\beta$ seconds, respectively. The number of mini sources in the 'on' state follows an underlying Markov chain, as shown in Figure 3.4. Denote by $Y_i(t)$ the sample bit rate of the video source at time t, and its mean and variance are $E[Y]$ and σ^2, respectively. The parameter set $(\alpha, \beta$ and $A)$ is obtained by matching the first and second moments of the composite Markov chain model and the measured statistics, i.e., the mean, variance and auto-covariance function (ACF) of the empirical data [52]:

$$E[Y] = MqA, \tag{3.1}$$

$$\sigma^2 = MA^2q(1-q), \tag{3.2}$$

$$\alpha + \beta = K. \tag{3.3}$$

The first two equations result from the conservation of the expectation and variance of the Markov chain and the data set, where M is the number of mini sources and $q = \alpha/(\alpha + \beta)$. The last equality in (3.3) comes from matching the ACF function of the Markov model to that of the data set. The ACF of the Markov chain is exponential in nature, where $\alpha + \beta$ is the so called 'time constant' of the exponential function. K is chosen by matching the ACF of the empirical data and the exponential decay function of $e^K(\tau)$.

To minimize the total distortion of the model, the algorithm to find K, known as the minimum mean square error (MMSE) algorithm, is defined by

$$\epsilon = \min \left(\sum_\tau |\rho(\tau) - e^K(\tau)|^2 \right), \tag{3.4}$$

where $\rho(\tau)$ is the ACF calculated from the data set defined as follows:

$$\rho(\tau) = \frac{1}{N-\tau} \sum_{m=1}^{N-\tau} \frac{(Y_m - E(Y))(Y_{m+\tau} - E(Y))}{\sigma^2}, \tag{3.5}$$

where τ is the lag of the two positions in the video sequence, σ^2 is the variance, and N is the total number of samples of the video source.

Since the sequences of video frames of many codecs (such as MPEG-2 and MPEG-4) have periodicity corresponding to the Group of Picture (GoP) pattern, the ACF of frame size, according to (3.5), is also periodic.[3] Therefore, it fails to match the aperiodic exponential decay function. Thus, the mini-source model uses

---

[3]For instance, with a typical MPEG video codec [61], a continuous video stream is sampled and quantized to form a sequence of frames as input to the encoder. After encoding, frames are emitted periodically comprising GoPs. Each GoP contains an I frame and a number of P and B frames. Typically, in a GoP, the I frame is large, and the B frames are small.

GoP size (or the average source rate during each GoP) as a reference. On the other hand, with the heavy tail property of video traffic, the ACF of GoP size will decay but not drop exponentially in terms of τ. However, if two GoPs are spaced far away, their considerable high correlation will not affect the buffer occupancy distribution significantly [62]. For instance, three GoPs already have a time interval of 1.2 s, which is a long enough duration compared with the time needed for queue states to reach equilibrium. Thus, in (3.4), τ is chosen from 0 to 3. The analytical and simulation results suggest that the cutoff value of $\tau = 3$ is a reasonable choice.

Because of its high compression and low data rate properties, we consider the MPEG-4 (H.264) video coding technology described in [61]. Even though the original work [54] suggested using 20 mini sources to model one video source, since H.264 has a much higher compression ratio than the source coding considered in [54], we recommend using a smaller number of mini sources to emulate the more bursty video sources.

3.2.2 A simple, two-level Markovian traffic model

As a GoP-level model, the mini-source model assumes constant data rate in each GoP duration, so it is not suitable for high speed networks [55]. As mentioned earlier, a typical MPEG video codec will encode a video stream into GoPs, and each GoP has an I frame and a number of P and B frames. With an MPEG codec, the generic GoP is $I\ B_1\ B_2\ P_1\ B_3\ B_4\ P_2\ B_5\ B_6\ P_3\ B_7\ B_8$. The first frame in each GoP is an I frame, which is intra-coded without reference to any other frames. The subsequent P frames are both intra-coded and inter-coded with respect to the previous P or I frames. The remaining B frames are also intra-coded and inter-coded, and they use both the previous and the following P or I frames as references. Since only I frames are encoded without exploiting temporal redundancy, typically an I frame has a much larger frame size than the B frames in each GoP. By averaging the traffic arrival rate within each GoP, the mini-source model underestimates the packet loss rate (PLR) of video sources over a bottleneck link with a small buffer size [63]. Since many popular video codecs use the GoP structure, and intra-GoP correlation plays an important role in the queueing behavior, the mini-source model has its limitations. Newer video models have been proposed to combine a Markov process with other stochastic processes to more accurately capture the video traffic characteristics. Many models reported in the literature [55, 57–60] focus on representing both short-range dependence (SRD) and long-range dependence (LRD) of a variable bit rate (VBR) video source or accurately matching the marginal distribution of the frame size. However, due to the complexity of these models, they are not easy to apply in the mathematical analysis of the queue behavior.

Since the video traffic models considered here are designed for network performance analysis and simulation studies, it is desirable to obtain a simple traffic model that can accurately estimate the statistics of the video traffic that have the greatest impact on the queue behavior. The following two-level Markov model, proposed in [64], partitions the video frames into different states with a simpler process, and both the inter-GoP and intra-GoP correlations are preserved.

GoP-level Markov chain

Video compression removes spatial and temporal redundancy within each video frame and within consecutive frames to enhance transmission efficiency. Hence, the frame size depends on the texture and motion complexity of the video content, or its spatial and temporal domain correlations. With lower texture complexity, the correlation in the spatial domain is high, and the resultant frame size is smaller. With lower motion complexity, the correlation in the temporal domain is high, so the resultant B and P frame size is smaller. In the spatial and temporal domains, we categorize the video stream into a number of levels, S and T, respectively. Thus, we can use $S \times T$ states to represent the correlations in both domains. Since the duration of a GoP is less than half a second, we assume that the spatial and temporal correlations of the video source in each GoP remain at the same level, or in the same state. Therefore, we can build a GoP-level discrete time Markov chain, with each state representing the temporal and spatial complexity of the video of the GoP.

The more states (larger S and T), the more accurate is the model, but at the cost of increased computational complexity in generating and applying the model. Experimental results show that choosing $S = 3$ and $T = 3$ makes a good trade-off between model accuracy and complexity. The three levels are denoted as low (L), medium (M) and high (H) complexity states.

The boundaries between states should be set appropriately, so that we can use a limited number of states to accurately capture the video statistics. By observing the experimental results with a few video streams, we note that evenly dividing the frames into each state can give reasonably good results, i.e.,

$$\Pr(S_i) = 1/N, \tag{3.6}$$

where S_i denotes one of the states, and $N = S \times T$ is the total number of states.

In the spatial domain, since only the I frames are independently intra-coded, the I frame size corresponds to the texture complexity of the entire GoP. In the temporal domain, the ratio of the first P frame size to the I frame size in the same GoP is used to indicate the temporal correlation, as explained below. The P_1 frame is both intra- and inter-coded with sole reference to the I frame. The frame size of P_1 is determined by $P_{1t} \times \Delta\varphi$, where $\Delta\varphi$ denotes the motion vectors from frame I to frame P_1 in the same GoP, and P_{1t} is the texture information contained in frame P_1, which is approximately the same as the texture information in the I frame of the same GoP. Hence, the ratio between the first P frame size and the I frame size in the same GoP ($\approx \Delta\varphi$) indicates the correlation in the temporal domain. A larger $\Delta\varphi$ indicates higher video motion speed, and lower correlation in the temporal domain. Thus, the ratio of P_1 to the I frame size in the same GoP can be used to determine the state in the temporal domain.

Combining the three states in the spatial domain with the three states in the temporal domain gives nine states for each GoP, as shown in Figure 3.5. For example, state *LH* denotes a GoP with low correlation in the spatial domain and high correlation in the temporal domain. Note that all states are connected. The transition probability between any two states is jointly determined by the temporal and spatial transition probabilities. Since the temporal and spatial processes are independent, the transition probability from state *LM* to state *MM*, for example, is

$$\Pr\{LM \to MM\} = \Pr_s\{L \to M\} \times \Pr_t\{M \to M\}, \tag{3.7}$$

where $\Pr_s\{L \to M\}$ and $\Pr_t\{M \to M\}$ denote the spatial transition probabilities from the L ('Low') state to the M ('Medium') state, and the temporal transition probability from the M ('Medium') state to the M ('Medium') state, respectively. Given a video trace, $\Pr_t$ and $\Pr_s$ can be obtained by counting the number of transitions between two states in the temporal domain and in the spatial domain, respectively.

Frame-level Markov chain

Since the GoP-level Markov chain cannot capture the burstiness of the traffic arrival rate within the GoP, we should further consider different types of frames inside each GoP and build a frame-level discrete time Markov chain.

The time step for the frame-level Markov model equals the duration of a video frame. As shown in Figure 3.5(b), with an MPEG codec, each state in the GoP-level model corresponds to a 12-step Markov chain at the frame level, which represents the 12 frames in the GoP. The state transitions within the GoP are deterministic. Each state corresponds to a different frame type with a different traffic arrival rate.

To use a limited number of states for a wide range of frame sizes, the critical issue is how to set the frame size for each state. The size of an I frame is determined by the spatial domain correlation only. Therefore, the I frame sizes in states *XL*, *XM* and *XH* are the same, where $X \in \{L, M, H\}$ in the spatial domain. Given a video sequence, a simple method to set the frame size of each I frame state is to average the size of all I frames belonging to that state:

$$\overline{I^X} = \sum_{I_j \in \{\text{state } XL, XM, XH\}} I_j / N^X, \quad \text{for } X = L, M, H, \tag{3.8}$$

where N^X is the total number of I frames in states *XL*, *XM* and *XH*, and I_j is the size of the I frame in the jth GoP. $\overline{I^X}$ in (3.8) is the average I frame size in states *XL*, *XM* and *XH*, which can be used to represent the I frame size in the corresponding states.

The B and P frame sizes, however, are determined by the combined spatial and temporal state. As discussed before, the ratio of P_1/I indicates the temporal correlation of each GoP. Thus, the average of P_1/I is used in the temporal domain state as

$$\alpha_{P_1}^K = \overline{P_1^K}/\overline{I^K},$$

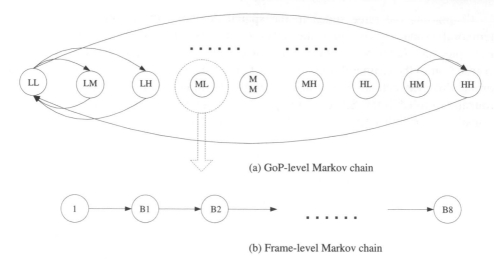

(a) GoP-level Markov chain

(b) Frame-level Markov chain

Figure 3.5 Two-level Markovian traffic model [64]. Reproduced by permission of ©2008 IEEE.

where $\overline{I^K}$ and $\overline{P_1^K}$ are calculated similar to (3.8), but in the temporal domain instead, i.e.,

$$\overline{I^K} = \sum_{I_i \in \{\text{state } LK, MK, HK\}} I_i / N^K, \tag{3.9}$$

where N^K is the total number of I frames in states LK, MK and HK, and I_i is the I frame size in the ith GoP. $\overline{I^K}$ in (3.9) is the average I frame size in states LK, MK and HK. Finally, the joint state decides the P_1 frame size as

$$\overline{P_1^{XK}} = \alpha_{P_1}^K \overline{I^X}. \tag{3.10}$$

By examining the intra-GoP correlation, the remaining P/B frame sizes in each state are generated using the same strategy as that for the P_1 size.

The correlation coefficient of the two sequences i and j is defined as

$$R(i, j) = \frac{CoV(i, j)}{\sqrt{CoV(i, i)CoV(j, j)}}. \tag{3.11}$$

We investigate some real MPEG-4 video traces. The lowest correlation coefficient between the B/P frames and P_1 in the same GoP is the last B frame B_8 and P_1, which is about 0.75 for the HD video 'From Mars to China' [65]. This indicates a high correlation between frames in the same GoP. Therefore, other P/B frame sizes are generated based on (3.10) as well:

$$\overline{T^{XK}} = \alpha_T^K \overline{I^X}, \quad \text{for } T \in \{B_1, B_2, \ldots, B_8, P_2, P_3\}. \tag{3.12}$$

In summary, the two-level Markov model is developed with two steps: first, categorize the GoPs into states in the temporal and spatial domains to build a GoP-level Markov chain, and determine the state transition probability matrix from the video trace; second, extend each GoP state to a 12-step[4] Markov chain in the frame level, and determine the frame sizes of each state based on the average values and linear relationships of the P/B and I frame sizes in the same state.

Remarks. The mini-source model, which is a GoP-level model, uses a coarser time scale. The time granularity of the mini-source model is equal to the duration of one GoP, e.g., 0.4 s for a 30 frames per second video trace, which is a relatively long time for high speed networks. This model smooths out the video traffic rate over the entire GoP duration, ignoring arrival rate variations within a GoP. For example, the higher I frame data rate is averaged by the low rate of the B and P frames in the same GoP. Thus, the mini-source model cannot capture the high loss rate of video packets when the buffer is not large enough to accommodate all packets from a large size I or P frame. The two-level model preserves both the inter-GoP and the intra-GoP correlations, which is more accurate for estimating network performance when the buffer size is small.

Another issue worth mentioning is the LRD property of video traffic. The challenge for a video model is to capture both the SRD and LRD properties of VBR video [66]. With just a constant rate modulated Markov process, the two Markovian models discussed here have an autocorrelation that drops exponentially. Thus, they cannot capture the LRD of a video trace. However, previous study showed that LRD is not a crucial property in determining the buffer behavior of VBR video sources [62]. The strong frame-to-frame correlation is more important, while the long-term form of the autocorrelation function, whether it is exponential over several frames or has some periodic component, has a negligible effect on the buffer statistics [56]. Simulation results in the following section with real video traces will also demonstrate that the buffer behavior of one or several multiplexed video sources can be accurately determined by the Markovian models without considering the LRD property of the video.

3.3   Performance Study of Video over Wired and Wireless Links

In this section, we focus on how to quantify the performance of multimedia traffic over wired and wireless links without end-to-end flow and congestion control. The performance study of multimedia traffic under flow and congestion control is much more complicated, and will be discussed in subsequent chapters, considering different types of control mechanisms and protocols.

---

[4]Twelve is the number of frames in each GoP, using an MPEG codec.

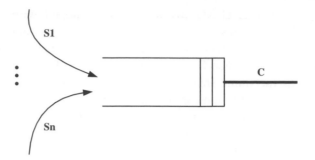

Figure 3.6 The multiplexed queue system.

The performance indexes for multimedia traffic are packet loss, delay and jitter, which are mainly related to the bottleneck queueing behavior. From the queueing theory's point of view, both the traffic arrival process and the packet service process affect the queueing behavior. The variation of traffic arrival is described by the traffic model, and the variation of service time of a wireless link is captured in the channel model. Thus, in [63], a uniform analytical framework to quantify the bottleneck queue behavior is introduced, which can incorporate the traffic models described in the previous sections of this chapter and the wireless channel models introduced in Chapter 2.

Consider the simple first-in–first-out (FIFO) queue shown in Figure 3.6. A number of video sources are multiplexed in the queue of the bottleneck link, which have a time varying service rate. The multiplexed video sources are described by the mini-source model, and the time varying channel has a data rate changing according to a Markov chain. Later, the analytical framework is extended to consider the case that the video traffic is multiplexed with other types of traffic. A similar approach can be applied to study the performance of multiplexed VoIP traffic with the birth–death traffic model, or that of video sources using the two-level Markovian model.

Combining the mini-source traffic model and the Markovian channel model, the system can be represented as a Markov chain with N states. In state $i \in \{1, 2, \ldots, N\}$, the traffic data rate is R_i and the link service rate is C_i. The generating matrix for the underlying (continuous time) Markov chain of the system is

$$\mathbf{B} = \begin{pmatrix} \mu_{11} & \mu_{12} & \cdots & \mu_{1N} \\ \mu_{21} & \mu_{22} & \cdots & \mu_{2N} \\ \vdots & \vdots & \vdots & \vdots \\ \mu_{N1} & \mu_{N2} & \cdots & \mu_{NN} \end{pmatrix}, \tag{3.13}$$

where μ_{kj}, $1 \leq k, j \leq N$ is the state transition rate from state k to state j, and $\mu_{kk} = -\sum_{j=1, j \neq k}^{N} \mu_{kj}$. Since the arrival rate changes with respect to the frame size and the duration of each frame is relatively long, the bottleneck buffer can quickly reach equilibrium when different size frames arrive [56]. We have the following theorem.

Theorem 1 *Let $F_i(x)$ denote the probability that the queue length is less than x, given that the system is in state i, $1 \le i \le N$. Let $\mathbf{F}(x)$ be the row vector*

$$[F_0(x) \; F_1(x) \; \cdots \; F_N(x)].$$

The equilibrium queue length distribution at the bottleneck link can be obtained by solving the system equation

$$\frac{d\mathbf{F}(x)}{dx}\mathbf{D} = \mathbf{F}(x)\mathbf{B}, \tag{3.14}$$

where

$$\mathbf{D} = \mathrm{diag}\{R_1 - C_1, \ldots, R_N - C_N\}$$

is an $N \times N$ diagonal rate matrix and $\mathbf{B}$ is a generating matrix given by (3.13).

Proof. Let $F_i(t, x)$ be the probability that the queue length at the bottleneck link does not exceed x at time t, when the system is in state i. The probability of the state transiting from state k to state j during a small time interval $\Delta t \to 0$ is $\mu_{kj}\Delta t$. Then, at time instant $(t + \Delta t)$,

$$F_i(t + \Delta t, x) = \sum_{k=1 \, (k \ne i)}^{N} \mu_{ki} \Delta t F_k(t, x) + \left(1 - \sum_{i=1 \, (i \ne j)}^{N} \mu_{ij}\right) \Delta t F_i(t, x - \Delta x) + o(t),$$

where $\Delta x = (R_i - C_i)\Delta t$ is the net gain of the buffer in the small interval t to $t + \Delta t$. Since $\Delta t \to 0$, the high order of Δt is ignored.

Adding $-F_i(t, x)$ to both sides and dividing both sides by Δt, we have

$$\frac{F_i(t + \Delta t, x) - F_i(t, x)}{\Delta t} = \sum_{k=1, \, k \ne i}^{N} \mu_{ki} F_k(t, x) - \sum_{i=1, \, i \ne j}^{N} \mu_{ij} F_i(t, x - \Delta x)$$

$$+ \frac{F_i(t, x - \Delta x) - F_i(t, x)}{\Delta x / (R_i - C_i)}.$$

In the limit as $\Delta t \to 0$ and $\Delta x \to 0$, we have

$$\frac{\partial F_i(t, x)}{\partial t} = \sum_{k=1, \, k \ne i}^{N} \mu_{ki} F_k(t, x) - \sum_{i=1, \, i \ne j}^{N} \mu_{ij} F_i(t, x) + (R_i - C_i)\frac{\partial F_i(t, x)}{\partial x}.$$

At equilibrium, $\partial F_i(t, x)/\partial t = 0$ and $F_i(t, x) = F_i(x)$, we have

$$(R_i - C_i)\frac{d F_i(x)}{dx} = \sum_{k=1, \, k \ne i}^{N} \mu_{ki} F_k(x) - \sum_{i=1, \, i \ne j}^{N} \mu_{ij} F_i(x). \tag{3.15}$$

Equation (3.15) is a set of scalar equations for $F_i(x)$, $1 \le i \le N$, which can be organized into the vector form shown in (3.14). □

According to [54], the solution of (3.14), i.e., the CDF of the queue length, is

$$F(x) = \sum_{k=1}^{N} F_k(x) = 1 + \sum_{i:Re[z_i < 0]} a_i \sum_{j=1}^{N} \phi_{ij} \exp(z_i x), \tag{3.16}$$

where $\{z_i\}$ and $\{\vec{\Phi}_i\}$ are the negative left eigenvalues and the corresponding eigenvectors of the matrix $\mathbf{BD}^{-1}$:

$$z_i \vec{\Phi}_i \mathbf{D} = \vec{\Phi}_i \mathbf{B}. \tag{3.17}$$

The coefficients α_i's can be obtained from the boundary conditions, i.e., $F_j(0) = 0$ for overload state j.[5] The survivor function $G(x)$, which represents the probability of buffer overflow, is the complementary distribution of $F(x)$:

$$G(x) = 1 - F(x) = - \sum_{i:Re[z_i < 0]} a_i \sum_{j=1}^{N} \phi_{ij} \exp(z_i x). \tag{3.18}$$

In summary, given the generating matrix $\mathbf{B}$ and the rate matrix $\mathbf{D}$, the queue length distribution can be obtained by solving (3.14). Then, from (3.18), the probability of buffer overflow can be predicted. In the following section, we determine the matrices $\mathbf{B}$ and $\mathbf{D}$ for different systems, from the simplest case with a wired link to a wireless link supporting heterogeneous traffic.

3.3.1   Transmission over a wired link

In the absence of multipath fading, the wireline channel is essentially time-invariant and the available bandwidth of a wired link is fairly constant. In the modeling of video transmission over a wired link, it is only necessary to model the video source. Thus, the generating matrix $\mathbf{B}$ in (3.14) is simply that of the video source. Let $\mathbf{B}_1$ and $\mathbf{B}_2$ be respectively the generating matrices of the mini source and the two-level Markov models introduced in the previous section. From Figure 3.4, we have

$$\mathbf{B}_1 = \begin{pmatrix} -M\alpha & M\alpha & 0 & \cdots \\ \beta & -(M-1)\alpha - \beta & (M-1)\alpha & \cdots \\ \vdots & \vdots & \vdots & \vdots \\ 0 & \cdots & \cdots & -M\beta \end{pmatrix},$$

and the corresponding rate matrix is

$$\mathbf{D}_1 = \mathrm{diag}\{0 - C, A - C, \ldots, M \times A - C\},$$

where M is the total number of mini sources.[6]

For the two-level discrete time Markov model, based on Figure 3.5, let $\mathbf{B}_2'$ be the state transition matrix for individual frames. Let $i = 1$ to 9 denote the states 'LL', 'LM', 'LH', 'ML', 'MM', 'MH', 'HL', 'HM' and 'HH' in

---

[5] An overload state is defined as a state in which the arrival rate exceeds the service rate.

[6] M is set to 8 for every H.264 video source, since the H.264 coded video is bursty and the simulation results show that using eight mini sources is suitable.

the model, respectively. Thus, if the states in $\mathbf{B}'_2$ are organized as $I^1, \ldots, I^9$, $P_1^1, \ldots, P_1^9, \ldots, P_3^9, \ldots, B_1^1, \ldots, B_1^9, \ldots, B_8^9$, $\mathbf{B}'_2$ has the following structure:

$$\mathbf{B}'_2 = \begin{pmatrix} \mathbf{O} & \mathbf{I} \\ \mathbf{BI} & \mathbf{O} \end{pmatrix},$$

where $\mathbf{I}$ is a 99×99 identity matrix which indicates deterministic transitions within each GoP. $\mathbf{B}'_2$ represents the expanded Markov chain in Figure 3.5, and $\mathbf{BI}$ is a 9×9 GoP-level matrix which is the transition matrix from the last frame (a B_8 frame) in the previous GoP to the I frame of the next GoP. Thus, $\mathbf{B}'_2$ has the dimension of 108 for every video source. Then, we can convert the discrete time Markov chain with state transition matrix $\mathbf{B}'_2$ to a continuous time Markov chain with generating matrix $\mathbf{B}_2 = (\mathbf{B}'_2 - \mathbf{I})/t$, where t is the frame duration. Consequently, the corresponding $\mathbf{D}_2$ matrix is

$$\mathbf{D}_2 = \mathrm{diag}\{R_1 - C, R_2 - C, \ldots, R_{108} - C\},$$

where R_i is the data rate generated in state i, which equals the ratio of the frame size of the state to the frame duration.

Given $\mathbf{B}$ and $\mathbf{D}$, the queue distribution can be obtained using the fluid flow approach. Pertaining to the QoS requirements of video traffic, including the delay bound and loss rate, we can choose an appropriate buffer size to bound the delay, and limit the number of connections to ensure the loss rate. For instance, given the delay budget in a network, the maximum tolerable queueing delay and queue length can be determined, and hence the buffer size. From (3.18), we can obtain the maximum number of connections that can be supported with a guaranteed loss rate due to buffer overflow.

3.3.2 Transmission over a wireless link

The wireless channel is time-variant and nonstationary so that the service time, i.e., the time to successfully transmit a packet over the wireless link, is time-varying. As introduced in Chapter 2, a FSMC can be used to model the time-varying service rate (the inverse of service time) of the wireless link.[7]

Let C_j denote the service rate of the wireless link in state j, and assume that the total number of states of the wireless link is N_l. We then have $\mathbf{C} = \mathrm{diag}\{C_1, C_2, \ldots, C_{N_l}\}$. The generating matrix of the underlying continuous time Markov process $\mathbf{B_c}$ is defined similar to (3.13) with the dimension of N_l. The wireless link service rate and the arrival from multimedia sources are two independent Markovian processes with the number of states N_l and N_s respectively. Let $\mathbf{B_s}$ be the generating matrix for the video source. The aggregated system can be represented by a Markov chain with $N_l \times N_s$ states. Let the coupled source and link system state

---

[7]A discrete time FSMC with state transition probability matrix $\mathbf{P}$ can be mapped to a continuous time FSMC. The state transition rate matrix $\mathbf{B}$ of the continuous time FSMC is obtained by $\mathbf{B} = (\mathbf{P} - \mathbf{I})/T$, where $\mathbf{I}$ is an identity matrix, and T is the duration of each slot in the discrete time FSMC model.

$s = (i, j)$ be ordered lexicographically:

$$(1, 1) \ (1, 2) \ \cdots \ (1, N_l) \ (2, 1) \ \cdots \ (N_s, N_l),$$

and, in the aggregated system, state $k = N_l(i - 1) + j$ corresponds to state (i, j), for $1 \le k \le N_l N_s$. As proved in [67], the generating matrix of the integrated chain can be obtained as

$$\mathbf{B} = \mathbf{B_s} \oplus \mathbf{B_c} \triangleq \mathbf{B_s} \otimes \mathbf{I}_{N_l} + \mathbf{I}_{N_s} \otimes \mathbf{B_c}. \qquad (3.19)$$

Here, $\oplus$ and $\otimes$ are the Kronecker sum and Kronecker product [67], respectively.[8] In (3.19), $\mathbf{I}_{N_l}$ and $\mathbf{I}_{N_s}$ are $N_l \times N_l$ and $N_s \times N_s$ identity matrices, respectively; thus $\mathbf{B}$ is a matrix with dimension $N_l N_s$. $\mathbf{D}$ is the diagonal matrix

$$\mathbf{D} = \mathbf{R_s} \oplus [-\mathbf{C}], \qquad (3.20)$$

where $\mathbf{R_s} = \mathrm{diag}\{R_1, R_2, \ldots, R_{N_s}\}$.

Given $\mathbf{B}$ and $\mathbf{D}$, the probability of buffer overflow can be obtained by solving (3.14). From the perspective of the bottleneck buffer, the output data rate can be viewed as a deduction on the instantaneous incoming traffic rate in each state.

The effective capacity [68] reflects the *stochastic bounded capacity* for a (or a number of) traffic source(s) accessing the buffer, conditioned on the loss probability P_L and the maximum buffer size x [69],

$$C_e = \frac{1}{\theta} z(\mathbf{B} + \theta \mathbf{D}),$$

where

$$\theta = \frac{1}{x} \ln(1/P_L),$$

and $z(\mathbf{A})$ represents the largest eigenvalue of matrix $\mathbf{A}$. High variance input traffic has a higher effective capacity, which leads to fewer connections that can be accommodated by a link. On the other hand, with highly variant output links, the effective link capacity will decrease, which also results in fewer admitted connections. This tendency is also verified by simulation results.

3.3.3 Multiplexing heterogeneous traffic with class-based queueing

When heterogeneous traffic (e.g., data, voice and video) is mixed in the network, the traffic characteristics and QoS constraints of different applications need to be considered. Data traffic has stringent loss requirement and prefers high throughput, while voice and video have stringent delay and jitter requirements. To ensure the QoS of heterogeneous traffic, we can deploy class-based queue (CBQ) management

---

[8]If $\mathbf{A} = \begin{bmatrix} a_{11} & \cdots & a_{1n} \\ \vdots & \ddots & \vdots \\ a_{m1} & \cdots & a_{mn} \end{bmatrix}$ and $\mathbf{B}$ are m-by-n and i-by-j matrices, respectively, $\mathbf{A} \otimes \mathbf{B} = \begin{bmatrix} a_{11}\mathbf{B} & \cdots & a_{1n}\mathbf{B} \\ \vdots & \ddots & \vdots \\ a_{m1}\mathbf{B} & \cdots & a_{mn}\mathbf{B} \end{bmatrix}$.

schemes in the bottleneck link. CBQ is a simple link-sharing approach that enables the gateway to distribute bandwidth on local links in response to local needs [70].

Note that unlike the generalized processor sharing (GPS) approach (e.g., weighted fair queue) that guarantees the long-term average bandwidth received by different classes, the CBQ scheme adopted here guarantees that each traffic class receives its allocated bandwidth over the relevant time interval. Thus, it can protect video/voice traffic from bursty data traffic. Even if the data traffic is idle for a long period, it cannot occupy more instantaneous bandwidth as compensation (which is not true using GPS), so the instantaneous bandwidth allocated to video/voice traffic is guaranteed. On the other hand, if any traffic class is idle, the other classes can occupy the total link bandwidth to achieve multiplexing gain. Without tracking the history of ongoing flows, CBQ is much simpler to implement than GPS-based schedulers. The different class queues are virtual queues for implementation. All packets are multiplexed into one queue with the condition that a number of packets belonging to one class are inserted in between packets belonging to the other classes. The maximum portion of packets belonging to each class is set according to the link-sharing parameters and the size of the packets. Thus, CBQ can easily be implemented in practice.

Consider the situation that video and data applications share the link with CBQ, where the data packets and video packets are classified into two classes.[9] In the absence of congestion, all packets are served FIFO. When congestion occurs (when the total arrival rate exceeds link capacity), the link-sharing scheduler of CBQ will rate-limit the over-limit class(es) to their assigned capacity. In this way, CBQ can prevent starvation of any class of traffic and ensure the QoS requirements. To implement CBQ, two virtual queues are deployed in the system: one for video and another to accommodate data. Each traffic class can only consume its own assigned bandwidth if the other virtual queue is not empty; otherwise, it can use all of the available link bandwidth.

Let p denote the portion of the service rate assigned to data, and $1 - p$ the service rate assigned to video when both queues are not empty. Let F_v and F_d denote the queue occupancy of video and data, respectively. The equations of F_v and F_d evolve as follows:

$$\begin{cases} \dfrac{dF_v}{dt} = \gamma^{(v)}(s_v) + pC - C & \text{if } F_d > 0, \\ \quad\ = \gamma^{(v)}(s_v) + \gamma^{(d)}(s_d) - C & \text{if } F_d = 0, \\[2mm] \dfrac{dF_d}{dt} = \gamma^{(d)}(s_d) + (1-p)C - C & \text{if } F_v > 0, \\ \quad\ = \gamma^{(d)}(s_d) + \gamma^{(v)}(s_v) - C & \text{if } F_v = 0, \end{cases} \qquad (3.21)$$

[9]Voice traffic can be represented as another separate class. Since the volume of voice traffic is much smaller than that of video or data traffic, its impact on the performance and admission region of data and video connections is limited, and therefore voice traffic is ignored here. Although we are assuming two classes of traffic here, the analysis presented in this section can easily be extended to any number of classes.

where the arrival rates of video and data are $\gamma^{(v)}(s_v)$ and $\gamma^{(d)}(s_d)$, respectively. $\mathbf{S} = (s_v, s_d)$ denotes the combined system state. Since the queue distribution derivations for video and data are similar, we study the video queue as an example.

The matrix $\mathbf{B_v}$ for the video traffic is decided by

$$\mathbf{B_v} = \mathbf{B_s^{(v)}} \oplus \mathbf{B_s^{(d)}},$$

where $\mathbf{B_s^{(v)}}$ and $\mathbf{B_s^{(d)}}$ are the generating matrices of the underlying Markov chain for video and data, respectively. Although $\mathbf{B_d}$ and $\mathbf{B_v}$ may be equivalent for the video and data queues, $\mathbf{D_d}$ and $\mathbf{D_v}$ are different. $\mathbf{D_v}$ is obtained as

$$\mathbf{D_v} = \mathbf{R_s} - C\mathbf{I} = \mathbf{R_v} \oplus \mathbf{R_{d'}} - C\mathbf{I}, \tag{3.22}$$

where $\mathbf{R_v}$ is the diagonal matrix with diagonal elements corresponding to the video arrival rate $\gamma^{(v)}(s^{(v)})$ of each state $s^{(v)}$ and $\mathbf{R_c}$ is the diagonal matrix with diagonal elements corresponding to the service rate. When the arrival rate of the data traffic $\gamma^{(d)}(s_d)$ is larger than its assigned bandwidth pC, the link-sharing scheduler in CBQ will limit the data traffic to its assigned bandwidth. Thus, at those states $\mathbf{S} = (s_v, s_d)$ with $\gamma^{(d)}(s_d) > pC$, the arrival rate of the data is pC instead of $\gamma^{(d)}(s_d)$ from the video traffic's point of view. $\mathbf{R_{d'}}$ is the diagonal matrix with the diagonal elements equal to the data arrival rate (from the video's viewpoint) $\min(pC, \gamma^{(d)}(s_d))$ of each state $s^{(d)}$. The coefficients a_i in the survivor function of (3.18) can be obtained from the boundary conditions, i.e., $F_o(0) = 0$ for $R(\mathbf{S}) > C$. Finally, we can obtain the packet loss ratio of the video packets.

From (3.21), buffer overflow can only occur in those states when $F_v > 0$ and the overall traffic arrival rate is greater than the total service rate. In equilibrium, when the queue is not filling, the probability of queue overflow is 0. The combined Markov process can be separated into two *virtual* states: overflow state Ω_o with $F_v > 0$ and underflow state Ω_u with $F_v = 0$. In state Ω_o, F_v is filling and video uses up all its assigned bandwidth. Overflow probability $\Pr(\Omega_o)$ equals the sum of the steady state probabilities of the states where the total arrival rate is larger than the capacity. In the overflow state, congestion will occur and CBQ will limit the over-limit traffic to its allocated bandwidth, so that we have

$$\Pr(F_v \geq x) = \Pr(F_v \geq x | \Omega_o)\Pr(\Omega_o).$$

In summary, given the buffer size, traffic models and bandwidth assignments, we can determine the loss rates of heterogeneous traffic sharing the same link with CBQ. Given the tolerable loss rates of different classes of traffic, we can further determine their admission regions. Finally, the analytical framework can be used to consider heterogeneous traffic over time-varying wireless links, using a similar approach by obtaining B and D according to (3.13) and (3.20).

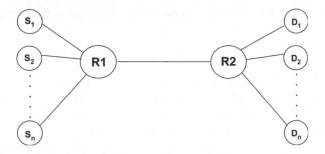

Figure 3.7 The network topology.

3.3.4   Simulation results

To verify the analytical framework, the analytical results are compared with simulation results for real video traces, using NS-2 [71].

Parameter settings

The network topology used in the simulations is shown in Figure 3.7. There are N video and/or data source and destination pairs. A source is connected to router R1 by a 100 Mbps wired link. The video stream is forwarded by the two routers R1 and R2 before arriving at the destination. All video and data traffic are encapsulated in 1000-byte UDP packets. All destination nodes are connected to router R2. The link between R1 and R2 is the bottleneck link under investigation. It would be a wired link or a time-varying wireless link. The bottleneck link buffer resides in R1. The routers use the simple Drop-Tail queue management scheme for solo video traffic, and CBQ is implemented for the case with heterogeneous traffic. The start time of any video source is randomly chosen to avoid synchronism among them.

Considering a UWB wireless link, its data rate followed a three-state Markov chain. The payload data rate is obtained by considering the physical layer data rate and the protocol overheads. The payload transmission time, T_{pay}, equals $payload/R_i$, and R_i can be 55 Mbps, 110 Mbps or 200 Mbps, depending on the current channel state. Using the parameters specified by the IEEE 802.15.3a as an example, the time to transmit a frame, T_{frame}, is

$$T_{frame} = T_a + \left(\frac{payload + RUI_h}{R_i} \right) + \left(\frac{PHY_h + MAC_h + HCS + FCS}{R_1} \right),$$

where T_a, RUI_h, PHY_h, MAC_h, HCS and FCS are the preamble time, RTP/UDP/IP headers, PHY header, MAC header, Header Check Sequence and Frame Check Sequence, respectively. The PHY/MAC overheads are transmitted at a base data rate

R_1 of 55 Mbps. The achievable throughput for the payload is then given as

$$C_i = \frac{T_{pay}R_i}{T_{frame} + T_g + T_{ACK} + 2SIFS},$$

where the guard time T_g, the ACK time T_{ACK}, and the Short InterFrame Space (SIFS) time are all equal to 10 μs, while all other overheads take the same values as in the standard [72, 73]. This leads to service rates of $C_1 = 40.3$ Mbps, $C_2 = 64.0$ Mbps and $C_3 = 88.7$ Mbps in states s_1, s_2 and s_3, respectively.

The wireless link is represented by a discrete time FSMC with time step duration T. The state transition probability of the wireless link can be obtained from measurement results or from channel modeling techniques introduced in Chapter 2. Here, we use the following matrix $\mathbf{P_c}$, where P_{ij} in the ith row and jth column is the transition probability from state i to state j:

$$\mathbf{P_c} = \begin{pmatrix} 0.1 & 0.9 & 0 \\ 0.1 & 0.2 & 0.7 \\ 0 & 0.1 & 0.9 \end{pmatrix}.$$

In the fluid model approach, we use a continuous time FSMC to describe the link capacity variation. The state transition rate matrix $\mathbf{B_c}$ of the continuous time FSMC is obtained by

$$\mathbf{B_c} = (\mathbf{P_c} - \mathbf{I})/T,$$

where μ_{ij} in the ith row and jth column is the transition rate from state i to state j, $\mathbf{I}$ is an identity matrix, and T is set to 20 ms unless otherwise specified.

Two HD video streams are chosen: 'From Mars to China' ('Mars') and 'Sony Digital HD Video Camera Demo' ('Sony'). Both are H.264 coded with 1920 × 1080 resolution [65]. The first is coded from an uncompressed raw video stream, while the latter comes from a WM9 format video file. 'Mars' has quantization parameters 28, 28 and 30 for the I, P and B frames, respectively, while the corresponding values for 'Sony' are 22, 22 and 24. The larger the quantization value, the higher is the compression ratio. With smaller quantization parameters, 'Sony' is coded with more details, e.g., its average I, P and B frame sizes are around 93.5 KB, 37.5 KB and 10.5 KB, respectively. The average data rate of 'Sony' is 5.8 Mbps, which is higher than that of 'Mars' with 4.8 Mbps. The peak data rate of 'Sony' (50.6 Mbps), however, is lower than that of 'Mars' (78.46 Mbps). Hence, 'Mars' is a more bursty video source. The ratio of peak to average rate is about 16 for 'Mars', which is twice that of 'Sony'. The simulations are repeated with different random seeds and each simulation runs for 30 minutes. The simulation results presented below are the averages of 20 runs.

Transmission over wired link

For a wired bottleneck link scenario, the single video flow case is addressed to validate the effectiveness of the video source models on network performance

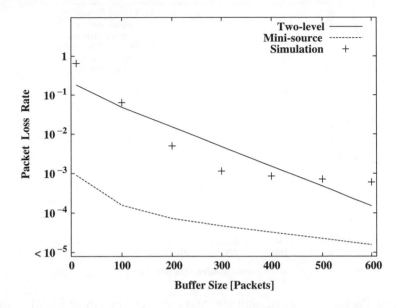

Figure 3.8 Packet loss rate of one 'Mars' connection with a 15 Mbps link [64]. Reproduced by permission of ©2008 IEEE.

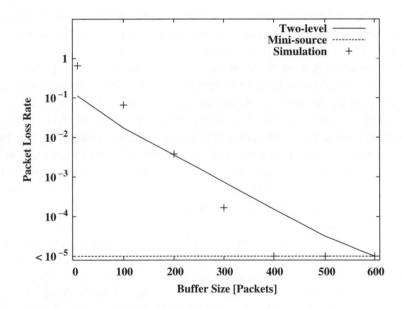

Figure 3.9 Packet loss rate of one 'Mars' connection with a 20 Mbps link [64]. Reproduced by permission of ©2008 IEEE.

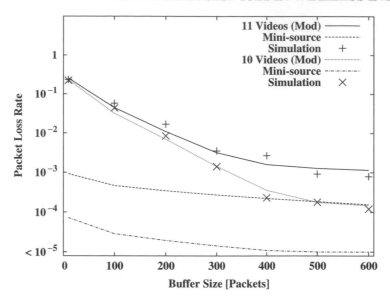

Figure 3.10 Packet loss rate of multiple 'Mars' connections in an 85 Mbps link [64].
Reproduced by permission of ©2008 IEEE.

estimation. The performance with the mini-source model and that with the two-level
Markovian model are investigated. The performance of multiple video flows, with
the two real video traces, is then evaluated.

Figures 3.8 and 3.9 show the PLR for one video source ('Mars') over a 15 Mbps
and a 20 Mbps bottleneck link, respectively. The analytical results with the two-
level Markovian model are close to the simulation results with the real video trace.
The analytical results with the mini-source model are also plotted in the figures.
As anticipated, using the mini-source model, the analysis underestimates the PLR,
especially when the buffer size is small, and it becomes more accurate when the
buffer size becomes larger. Since the mini-source model assumes a constant arrival
rate within a GoP duration, the results with the mini-source model can also be used
to predict the performance if we deploy a traffic shaping technology to smooth the
video traffic within a GoP.

Figures 3.10 and 3.11 show respectively the performance of multiple video
sources of 'Mars' and 'Sony' sharing an 85 Mbps link. We use the two-level traffic
model to generate synthetic video traces in the simulations, and compare the results
with the simulations using the real video traces. As shown in the figure, the resultant
packet loss rates match very well. Thus, the two-level traffic model is an effective tool
for fast simulation purposes. The analytical results using the mini-source model can
still be used to quantify the performance and the admission region when the buffer
size is large.

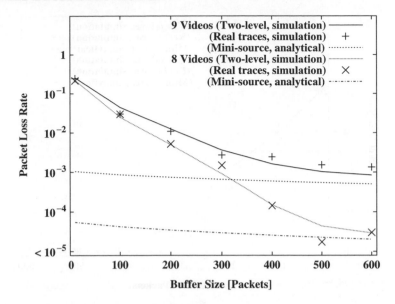

Figure 3.11 Packet loss rate of multiple 'Sony' connections in an 85 Mbps link [64]. Reproduced by permission of ©2008 IEEE.

Notice that the accuracy of the two-level model is at the cost of scalability. The number of states for one video source is 108, which is significant. Nevertheless, the matrix is sparse and simple to handle mathematically in terms of calculating eigenvalues and eigenvectors. If the mini-source model uses m mini sources for each video, and there are n sources sharing the link, the aggregated video traffic is an $nm + 1$-state birth–death Markov process. However, with the two-level Markov model considering intra-GoP correlations, due to the fully connected Markov chain, the number of states is 108^n. Hence the mini-source model is less accurate but more scalable for the analytical approach. In summary, it is recommended to use the mini-source model to analyze network performance with sufficient buffer size, and to use the two-level Markovian traffic model for analysis considering small buffer size or to generate synthetic video traffic for simulations.

Transmission over wireless link

As shown in Figure 3.12, for a wireless link with an average data rate of 85 Mbps, when a buffer size is greater than 200 packets, the PLR drops below 10^{-4} when admitting six video sources, while 400 packets are needed to admit seven videos. Comparing these results with those for the 85 Mbps wired link, the time-varying wireless link can admit far fewer video sources. This demonstrates that if we simply use the average data rate of a time-varying link to calculate the admission region, the results will be too optimistic.

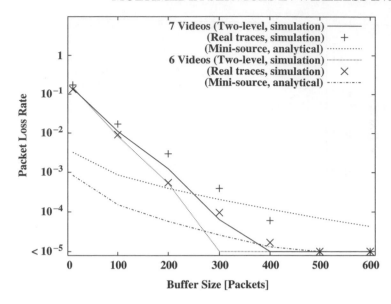

Figure 3.12 Packet loss rate of multiple 'Mars' connections in a wireless link.

The gap between the mini source analytical results and the simulation results also suggests the usefulness of traffic shaping or smoothing. If we can smooth the video sources over GoPs, the buffer size needed to ensure a certain PLR will be significantly reduced, and more video sources can be supported.

Heterogeneous traffic

Next, we study the performance of video traffic competing with 5 Mbps CBR data traffic, and with bursty on/off data traffic with a peak rate of 5 Mbps, respectively. For the on/off data source, the average duration of the off period is ten times that of the on period. The average on period is set to be 55.605 s according to the Internet measurements in [74]. A 'virtual' CBQ is deployed at the router R1, as shown in Figure 3.13. There are separate virtual queues for different traffic classes. The video buffer ranges from 400 packets to 1100 packets, and the data buffer size is set to the closest integer to $(1/p - 1)$ times the video buffer size.

Figures 3.14 and 3.15 show the analytical and simulation results for PLR vs. buffer size, with a CBR data source and an on/off data source, respectively. For the CBQ, 5% of the link bandwidth is reserved for data, and the remaining bandwidth is used to support video traffic. The buffer size shown in the horizontal axis denotes the total buffer for both video and data traffic.

In Figure 3.14, both the analytical and the simulation results show that, if one data source keeps transferring at a CBR of 5 Mbps, to ensure the PLR is less than 10^{-4}, a maximum of ten video connections can barely be supported with a buffer

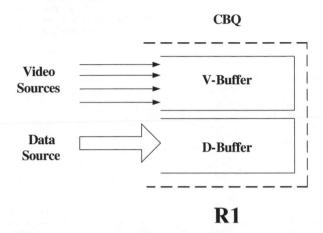

Figure 3.13 Class-based queueing resides at the router R1.

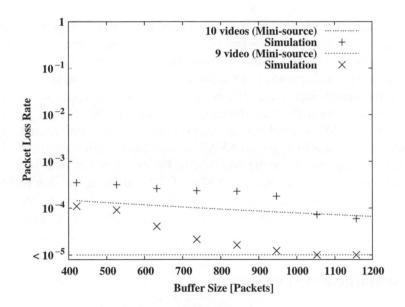

Figure 3.14 Performance of video (competing with constant bit rate data) in an 85 Mbps wired link [63]. Reproduced by permission of ©2007 IEEE.

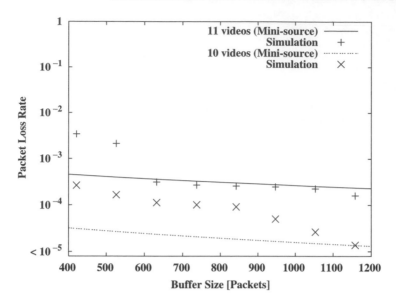

Figure 3.15 Performance of video (competing with on/off data) in an 85 Mbps wired link [63]. Reproduced by permission of ©2007 IEEE.

size larger than 1000 packets. Nine video connections can be supported with a safe margin. The fluid model analysis provides a good approximation of PLR when the buffer size is large, and the admission region derived is meaningful.

Typically, data traffic tends to be more bursty than CBR, and we use the on/off model for the bursty data traffic. Given an appropriate scheduling mechanism like CBQ, video and data traffic can efficiently share the link bandwidth. As shown in Figure 3.15, an 85 Mbps wired link can support ten video connections and one on/off data connection. Therefore, for an 85 Mbps wired link with CBQ, the admission region is ten video sources in the presence of the on/off bursty data source. In the worst case where the data source is a 5 Mbps CBR one, the admission region drops to nine video connections. In summary, if we allocate resources for different traffic classes appropriately using CBQ, both video and data traffic can be supported with satisfactory QoS.

3.4   Scalable Source Coding

For delay-sensitive multimedia applications over best-effort and highly dynamic networks, scalable source coding schemes with an error concealment property have been proposed and are anticipated to be widely deployed [13–18, 75]. For instance, in the Amendment of MPEG-4, a Fine Granularity Scalability (FGS) video-coding technique is developed for streaming video over the Internet [14, 15]. With the FGS technique, a video sequence is coded into a non-scalable base layer and a scalable enhancement layer. A certain degree of packet losses in the enhancement layer is tolerable; but packet losses in the base layer may result in severe distortion of the video, and that is intolerable. Therefore, FGS is most suitable for networks with differentiated services.

Another approach is the non-layered Multiple Description (MD) coding [13]. An MD coder partitions the source data into several sets and then compresses them independently to produce descriptions. The quality of reconstructed multimedia can be improved when more descriptions are received, and the MD decoder can conceal packet losses up to a certain degree. Since MD coding does not require differentiated services, it is very attractive for multimedia applications over networks without service differentiation.

With scalable source coding schemes, the video traffic volume can adapt to the available network resources, according to the flow and congestion control deployed in the transport layer. Performance study of flow and congestion controlled scalable video traffic over wired and wireless networks is much more complicated, and will be treated in Chapters 6 and 7.

3.5 Summary

For queueing performance of multimedia traffic over wired and wireless links, both traffic and link characteristics play important roles. In this chapter, we have studied the modeling of both VoIP traffic and video traffic. We then introduced an analytical framework to quantify the queue distribution for multimedia flows over wired or wireless links, with both single class traffic and heterogeneous traffic, where the multimedia sources do not change their data rates according to network conditions. In the following chapters, we will study the flow and congestion control mechanisms in the Internet, and investigate the end-to-end performance of adaptive multimedia traffic under flow and congestion control.

3.6 Problems

1. A voice source is modeled as an 'on/off' source. The average duration of the active period is T_a s and the average duration of the silent period is T_s s. During the active period, a voice source will generate traffic at rate $A = 64$ Kbps, and during the silent period, no traffic is generated.

    (a) What is the probability that a voice is in the 'on' state?

    (b) If there are K voice sources sharing a link, what is the average data rate?

    (c) If the link capacity is C Kbps and K voice sources share the link, what is the percentage of time that the traffic arrival rate exceeds the link data rate?

(d) Given $T_a = 0.5$ s, $T_s = 0.5$ s, $A = 64$ Kbps, what is the minimal C needed to support 20 voice sources such that the percentage of time that the traffic arrival rate exceeds the link data rate is less than 0.05?

(e) Assuming that the voice source generates traffic at 5 Kbps during the active period, repeat parts (a)–(d).

2. Given $\mathbf{D}$ and $\mathbf{B}$, obtain $\mathbf{F}(x)$ of the following system:

$$\frac{d\mathbf{F}(x)}{dx}\mathbf{D} = \mathbf{F}(x)\mathbf{B}.$$

3. For the mini-source video model, the average residence times in the 'off' state and 'on' state are $1/\alpha$ and $1/\beta$ seconds, respectively. Derive the following equations given in Subsection 3.2.1:

$$E[Y] = MqA,$$
$$\sigma^2 = MA^2 q(1 - q),$$

where $E[Y]$ and σ^2 are the mean and variance of the video source data rate per GoP, A is the data rate (per GoP) of each mini source during the 'on' state, M is the number of mini sources and $q = \alpha/(\alpha + \beta)$.

4. The bit rate of a compressed video sequence can be represented by the following autoregressive model: $\lambda(n) = a\lambda(n - 1) + bw(n)$, where $|a| < 1$ and $w(n)$ is a stationary white Gaussian noise (i.i.d.) process with unit variance and average value $E(w) = \eta$.

Show that the average bit rate is given by $E(\lambda) = b\eta/(1 - a)$, $|a| < 1$.

Note that the autocovariance function decays exponentially as

$$C(n) = \sigma_\lambda^2 a^n, \quad n \geq 0,$$

where $\sigma_\lambda^2 = C(0) = E(\lambda^2) - E^2(\lambda) = b^2/(1 - a^2)$.

5. Let three 'on/off' flows share a link with capacity C. A_i is the data rate of flow i during its 'on' state. The average durations for 'on' and 'off' states for flow i are T_{Ai} and T_{Si}, respectively. The link deploys CBQ to support them. Let p_i and F_i ($i = 1, 2, 3$) be the portions of service rate assigned to the three flows and the queue lengths for each of them, respectively.

(a) Find dF_i/dt, for $i = 1, 2, 3$.

(b) Given $C = 1$ Mbps, $p_i = 1/3$, $A_i = 400$ Kbps, $T_{Ai} = 0.4$ s, and $T_{Si} = 0.6$ s, for $i = 1, 2, 3$, find the queue length distribution.

6. A mini-source model is used to model five video sources multiplexed at an access buffer. The frame rate of the video source is 30 frames per second, and each GoP has 12 frames. The average GoP size of each source is 225 KB. The standard deviation of GoP size is $\sigma = 100$KB. The autocovariance time constant is 2.5 s. The source output is converted to 1460-byte packets before transmission to the buffer. The buffer in turn transmits into the network at a rate of 2400 packets/s. Use eight mini sources to model each video source. Apply the analytical framework in Section 3.3 to obtain the survivor function $G(x)$.

7. Download the video trace file of H.264 encoded 'From Mars to China' from http://www.ece.uvic.ca/~cai/verbose.data.

   (a) Construct a mini-source model (using eight mini sources) for the video source.

   (b) The video traffic goes through a wired bottleneck link of 20 Mbps. Using the mini-source model, apply the analytical framework in Section 3.3 to obtain the survivor function $G(x)$.

   (c) Build a two-level Markovian model for the video source.

   (d) Let the video traffic go through a wired link of 20 Mbps. Using the two-level Markovian model, apply the analytical framework in Section 3.3 to obtain the survivor function $G(x)$.

   (e) Traffic from eight video sources goes through a wireless bottleneck link. The wireless link can be modeled as a three-state Markov chain with service rates $C_1 = 40.3$ Mbps, $C_2 = 64.0$ Mbps and $C_3 = 88.7$ Mbps in states s_1, s_2 and s_3, respectively. The state transition rates (per second) of the wireless link are given in the matrix $\mathbf{B_c}$ below, where μ_{ij} in the ith row and jth column is the transition rate from state i to state j.

   $$\mathbf{B_c} = \begin{pmatrix} -18 & 18 & 0 \\ 2 & -16 & 14 \\ 0 & 2 & -2 \end{pmatrix}.$$

   Using the mini-source model, apply the analytical framework in Section 3.3 to obtain the survivor function $G(x)$.

   (f) Traffic from a video source is multiplexed with 'on/off' data traffic in a wired bottleneck link of 20 Mbps, where the access buffer deploys a simple FIFO queue. The average 'on' and 'off' periods of the data source are set to 55.605 sec and 556.05 s, respectively. During the 'on' period, the data source generates 5 Mbps traffic. Using the mini-source model, obtain the survivor function $G(x)$.

   (g) A CBQ is deployed in the access buffer, where 5% of the link bandwidth is reserved for data, and the remaining bandwidth is used to support the video traffic. Repeat part (f).

8. Use the following traffic-shaping technology to smooth out the MPEG-4 video source: the sender buffers packets of a whole GoP and sends the packets at constant rate over the duration of a GoP. Let the smoothed video traffic go through a wired link of R Mbps. Explain your approach and describe the steps to obtain the survivor function $G(x)$.

9. Compare the advantages and disadvantages of using a CBQ or a GPS scheduler to support mixed non-elastic video traffic and TCP-controlled data traffic in a link.

4

AIMD Congestion Control

4.1 Introduction

With the universal adoption of the Internet as a global information transport infrastructure, besides the traditional Web applications, more and more highly demanding and media-rich applications continue to emerge and become very popular. Although the HTTP-based Web traffic still dominates the Internet, it is well anticipated that new multimedia applications will claim a large percentage of the Internet traffic mixture in the future.

The success of the Internet in the past has relied heavily on the self-regulated TCP in its transport layer. A TCP sender does not assume any explicit knowledge of the internal structure of the network and other users. The TCP sender uses a congestion window (*cwnd*) to probe network conditions. Packet losses, presumably due to network congestion, are used as the basis for generating congestion indicators. When a congestion indicator is captured, the TCP sender reduces its *cwnd* by half, or sets the *cwnd* to the initial window size for severe congestion. Otherwise, the TCP sender probes for unused bandwidth aggressively by enlarging the *cwnd* by one segment per round-trip time (*rtt*).

Although it has been shown that TCP congestion control is very successful for bulk data transfer, the *increase-by-one* or *decrease-by-half* strategy produces a highly fluctuating sending rate, which is undesirable for many multimedia applications. Since TCP-transported applications are dominant in the Internet, it is crucial to have compatible traffic regulations for non-TCP applications. These regulations, or congestion control mechanisms, are expected to achieve the following objectives: (a) different classes of multimedia applications can coexist and behave properly; (b) these multimedia applications can share the network resources appropriately with ordinary TCP-transported ones. We refer to these regulations as *TCP-friendly congestion control for non-TCP-transported multimedia applications*. In addition to

Multimedia Services in Wireless Internet: Modeling and Analysis   Lin Cai, Xuemin Shen and Jon W. Mark
© 2009 John Wiley & Sons, Ltd

the fairness and TCP-friendliness issues, any new congestion control scheme should also (a) have the ability to maintain network stability by promptly responding to congestion and be cooperative with other flows, (b) utilize the network resources (e.g., link bandwidth) efficiently, (c) be capable of providing better QoS (e.g., a smoothed flow and bounded latency for multimedia playback applications), and (d) be simple to implement, compatible with the legacy and scalable for incremental deployment.

There are two paradigms of TCP-friendly congestion control schemes proposed in the literature: *equation-based rate control* and *generic AIMD-based window control*. For the equation-based approach, several analytical models [76,77] are used to obtain the long-term TCP throughput as a response function of the measured packet[1] loss event rate (p), *rtt*, maximum segment size (MSS), and initial timeout value (T_0). Equation-based rate control and its performance will be discussed in Chapter 7.

The generic AIMD-based congestion control inherits the same congestion control mechanism as is embedded in TCP. Instead of the increase-by-one or decrease-by-half strategy, the AIMD sender increases its *cwnd* by α segments additively when no congestion is sensed; otherwise, it multiplicatively decreases the *cwnd* to β times its previous value [78]. AIMD(α, β) is a general additive increase and multiplicative decrease congestion control protocol, and TCP is a special case of AIMD, with $\alpha = 1$ and $\beta = 0.5$. With an appropriate pair of (α, β), it is possible to have a smoothed sending rate and, at the same time, to be TCP-friendly. There are other window-based control variants proposed in the literature, e.g., Rate Adaptation Protocol (RAP) [79] explores the inter-packet gap (IPG) to exercise congestion control. If no congestion is sensed, IPG is reduced additively; otherwise, IPG is doubled multiplicatively. This technique can also be applied to inter-acknowledgment spacing-based approaches [80].

We focus on the generic AIMD-based congestion control algorithms and their α and β parameters. Since the AIMD algorithms inherit the same principles as are embedded in TCP, they can be incrementally implemented and deployed on a large scale for non-TCP applications. To pursue a deeper understanding of the intrinsic properties and the performance of AIMD, the competitive behaviors among TCP and AIMD-controlled flows should be investigated; this is precisely the treatment of this chapter. The *necessary* and *sufficient* TCP-friendly condition for a family of AIMD parameters has been derived in [81]. It has been demonstrated that the condition is valid for any number of multi-class AIMD and TCP flows. Based on the analysis of its built-in properties of effectiveness and responsiveness, and the discussion of practical implications, a dynamic TCP-friendly AIMD (DTAIMD) algorithm, as proposed in [81], is described. The efficacy and performance of the TCP-friendliness condition and the DTAIMD algorithm are analyzed, and extensive performance studies are presented.

---

[1] The terms network *packet*s and transport *segment*s are used interchangeably in this text since transport layer protocols can negotiate for the maximum segment size on their connection establishment to avoid IP fragmentation.

The remainder of this chapter is organized as follows. In Section 4.2, we introduce the AIMD protocol, which is designed to regulate multimedia applications [82]. In Section 4.3, we introduce the models for one TCP and one AIMD, and for multiple-class AIMD flows, which are used to obtain the *necessary* and *sufficient* TCP-friendly condition [81]. The effectiveness and responsiveness of the AIMD algorithms and the practical implications are also discussed. The DTAIMD-based congestion control for multimedia applications, including service differentiation and low or high multiplexing scenarios, are then studied in Section 4.5. Performance studies are presented in Section 4.6, which gives the simulation results for flow fairness, link utilization, convergent speed, service differentiation and QoS provisioning of DTAIMD-based multimedia applications. Section 4.7 summarizes the main points presented in this chapter.

4.2   AIMD Protocol Overview

Similar to TCP, the AIMD protocol uses a window-based flow and congestion control mechanism. The AIMD sender sends packets whose sequence numbers are in the range of the sender window, which is sliding; the AIMD receiver sends acknowledgments (*ack*s) for correctly received packets. The sender window size, W, is determined and adjusted according to the flow and congestion control mechanism.

In the following, we focus attention on the main protocol components: the acknowledgment scheme, and the flow and congestion control mechanisms. Then, we discuss the advantage of the window-based AIMD congestion control mechanism. For other protocol details, the reader is referred to the specification of DCCP-2 [83].

4.2.1   Acknowledgment scheme

Since the AIMD sender is not obligated to retransmit corrupted or lost packets, cumulative acknowledgment, as used in TCP, is not applicable. The acknowledgment scheme of DCCP-2 requires the sender to acknowledge the receiver's acknowledgments, which is too complex to implement. Here, we present a simple and applicable acknowledgment scheme for the AIMD protocol.

The AIMD *ack* has two fields: a 24-bit acknowledgment number, ack_a, and a Selective acknowledgment Vector (SackVec) with negotiable length, as shown in Figure 4.1. ack_a identifies the packet with a valid sequence number (in circular sequence space) received from the sender; the ith bit in SackVec indicates the status (whether or not a packet has been received) of the packet with sequence number $(ack_a - i)$. The length of SackVec is denoted as $|SackVec|$.

With SackVec, the AIMD *ack*s have some redundancy and can tolerate occasional losses of *ack*s. The AIMD receiver can send one *ack* for several (up to $1 + |SackVec|$) packets received if the bandwidth consumption of *ack*s is of great concern. For a packet not being indicated in any *ack* within a certain time interval, the sender

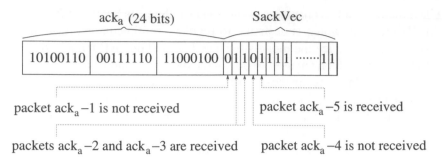

Figure 4.1 AIMD *ack*.

assumes that it is lost (which is a timeout event). Thus, the AIMD sender is not required to acknowledge *ack*s, and the receiver does not retransmit *ack*s.

4.2.2   Flow and congestion control

The AIMD receiver maintains a buffer and a window (*rwnd*). To avoid a fast sender over-running a slow receiver, the AIMD receiver advertises the amount of available buffer for the connection, and the AIMD sender uses the receiver's advertised window (*rwnd*) to bound the amount of in-flight packets.

AIMD receiver   The algorithm used at the AIMD receiver side is shown in Figure 4.2. Let $Rwnd_{left}$ and $Rwnd_{right}$ denote the left edge and right edge of the receiver window, respectively.

When the AIMD receiver receives a packet with sequence number s (in circular sequence space[2]):

1. If s is to the left of $Rwnd_{left}$ (i.e., $s < Rwnd_{left}$), the packet is discarded without any *ack*.

2. If s lies within the receiver's window ($Rwnd_{left} \leq s \leq Rwnd_{right}$), the received packet is buffered, and the receiver sends an *ack* which shows that the packet s is received and also indicates the status of the packets with sequence numbers between $s - |SackVec|$ and $s - 1$ in its SackVec.

3. If s is larger than $Rwnd_{right}$, the packet is buffered, and the receiver sends an *ack* to the receiver and advances $Rwnd_{right}$ to s (if $Rwnd_{left} < s - rwnd + 1$, $Rwnd_{left}$ is advanced to $s - rwnd + 1$). Therefore, the buffered packets always have the *largest* sequence numbers, which is desired for delay-sensitive multimedia applications.

---

[2]To compare two sequence numbers a and b in circular sequence space of 2^{24}, if $0 < b - a < 2^{23}$ or $b - a < -2^{23}$, b is larger than a; if $0 < a - b < 2^{23}$ or $a - b < -2^{23}$, b is smaller than a.

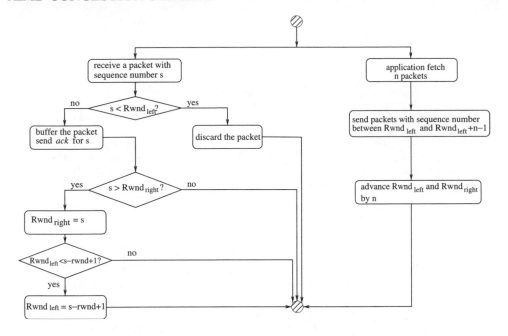

Figure 4.2 AIMD receiver [82]. Reproduced by permission of ©2006 IEEE.

When the multimedia application fetches n packets from the AIMD receiver buffer, the packets with sequence numbers between $Rwnd_{left}$ and $Rwnd_{left} + n - 1$ are delivered to the application, and $Rwnd_{left}$ and $Rwnd_{right}$ are advanced by n.

The congestion control mechanism used in the AIMD protocol is similar to that in TCP-SACK [84], except that a pair of parameters, (α, β), is used. To probe for available bandwidth and respond to network congestion, the AIMD sender uses a *cwnd* to control the sending rate. The actual size of the sender window (W) is the minimum of *cwnd* and *rwnd*.

The *cwnd* evolves in three phases. Initially, after a timeout, or being idle for a while, *cwnd* is set to a small initial window (IW), and it is doubled after each *rtt*, which is the *slow start* phase. Slow start threshold $(ssthresh)$ is set to reflect the estimated available bandwidth. When *cwnd* exceeds *ssthresh*, *cwnd* is additively increased by α packets per *rtt*, which is the *congestion avoidance* phase, until eventually congestion occurs. Severe congestion is indicated by timeout, which forces the AIMD sender to set *cwnd* to *IW* and *ssthresh* to one-half of the current window size (the minimum of *cwnd* and *rwnd*, but at least two segments), followed by slow start. Moderate congestion is indicated when the AIMD *acks* show that one packet is not received and three or more packets with a larger sequence number are received, and the former packet is assumed lost. When this occurs, *cwnd* is reduced by a factor of β (or more precisely, reduced to β times the number of currently outstanding packets) and $ssthresh = cwnd$, which is the *exponential backoff* phase.

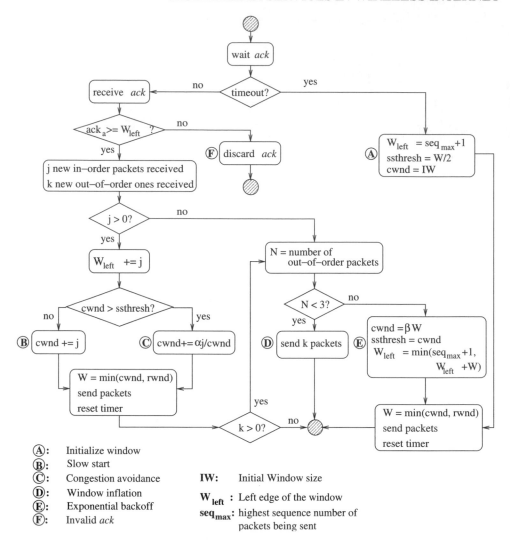

Figure 4.3 AIMD sender [82]. Reproduced by permission of ©2006 IEEE.

Similar to TCP, the AIMD sender will backoff only once for all packet losses within one window.

The flow and congestion control algorithm at the AIMD sender side is shown in Figure 4.3. Specifically, if timeout occurs, the AIMD sender re-initializes the *cwnd* and W, advances the left edge of the sender window (W_{left}), and sends W new packets (event **A** in the figure); if the current *cwnd* is 1, the new *cwnd* is still 1, but the inter-packet spacing is doubled such that the sending rate is halved. If an *ack* is received and it indicates that j packets with sequence numbers $W_{left}, \ldots, W_{left} + j - 1$ are received successfully, *cwnd* is enlarged by j if in the slow start phase (event **B**), or enlarged by $\alpha j / cwnd$ if in the congestion avoidance phase (event **C**). If the *ack* indicates that k out-of-order packets with sequence numbers larger than W_{left} are received, the window is inflated by k, and k new packets are sent if the number of out-of-order packets is less than 3 (event **D**); otherwise, the window is deflated, *cwnd* is decreased by a ratio of β, and W_{left} is advanced to guarantee that the window is reduced only once per *rtt* (event **E**). If ack_a is less than W_{left}, the sender discards the *ack* (event **F**). To avoid a burst of packets being pumped into the network, the inter-packet spacing[3] is set to no less than rtt / W when the sender is allowed to send more than one packet at a certain time.

4.2.3   Advantages of the window-based AIMD mechanism

The AIMD congestion control mechanism has the following advantages: (a) the success of the Internet over the past two decades has demonstrated the effectiveness and efficiency of the AIMD mechanism; (b) in general, anything slower than exponential backoff cannot guarantee network stability when the end-systems have no complete knowledge of the global traffic [85]; (c) AIMD is the *only* binomial control with monotonic convergence to fairness [86]; and d) the increase rate and decrease ratio (α, β) pair can be flexibly chosen to provide a wide variety of QoS to various multimedia applications.

In addition, window-based protocols have the *acknowledgment self-clocking* property, which is particularly useful to regulate traffic over time-varying wireless links. When the wireless channel is in a bad condition temporarily, the AIMD sender can immediately reduce the sending rate since the *ack*s are delayed due to low link throughput. Once the channel condition becomes better, the *ack*s can arrive at the sender at a faster pace, and the sending rate will be increased accordingly. Therefore, the queue length at the wireless link is always constrained by the window size.

4.3   TCP-friendly AIMD Parameters

AIMD has an identical slow start process initially or after a timeout to that of TCP. However, in congestion avoidance and exponential backoff phases, instead of using TCP's increase-by-one or decrease-by-half strategy, AIMD adjusts its *cwnd* according to two parameters: α and β. Therefore, we need to examine only the competitive behaviors between TCP and AIMD in the congestion avoidance and exponential backoff phases. Using the window-based congestion control scheme, an AIMD sender increases its *cwnd* by α segments per *rtt* to probe for unused bandwidth when there is no congestion, and decreases the *cwnd* to β times its previous value if a congestion indicator is captured. For easy reference, we list the notation used in this chapter in Table 4.1.

---

[3]The packet spacing technology has been deployed in RAP [79] and TCP rate control [80].

Table 4.1   Notation for Chapter 4.

Symbol	Description
(α, β)	AIMD parameter pair, the increase rate and decrease ratio of the *cwnd* in the congestion avoidance and exponential backoff phases
$W_A(t)$	Window size of the AIMD flow at time t
$W_T(t)$	Window size of the TCP flow at time t
$\overline{W}$	Average window size in steady state
t_i	The time instance that the flows enter the overload region ith times
Δt_i	$t_{i+1} - t_i$
t_*	The time instance that the flows enter the overload region in steady state
τ_0	Deterministic round-trip time
R	Link capacity, in terms of the number of packets being transmitted per τ_0
η	Link utilization
$c(\alpha, \beta)$	Convergent speed of AIMD(α, β) flows
$c(\alpha_1, \beta_1; \alpha_2, \beta_2)$	Convergent speed of AIMD flows with (α_1, β_1) pair and (α_2, β_2) pair
C	Consumption rate of multimedia playback applications
B_f	Volume of data stored in the receiver buffer during the filling phase
B_d	Volume of data drawn from the receiver buffer during the draining phase

4.3.1   One TCP and one AIMD flow

Let one TCP flow and one AIMD(α, β) flow share a link. Assume that they have the same *rtt* and packet size (the effect of different *rtt* and packet size will be discussed in Subsection 4.3.3). *cwnd* is calculated in the number of full size packets. In this section, *rtt* is assumed constant for all flows, time is measured in units of *rtt*, and the sender window size is determined by the *cwnd* alone (i.e., *rwnd* is always larger than *cwnd*). The description of the effect of *rwnd* is deferred to Chapter 6. The link has a capacity of R packets per *rtt*.

Figure 4.4(a) shows the *cwnd* trajectories of the two flows. Let $W_A(t)$ and $W_T(t)$ denote respectively the window size (equal to *cwnd*) of the AIMD flow and that of the TCP flow at time t. When $W_T(t) + W_A(t) < R$, referred to as the *underload* region

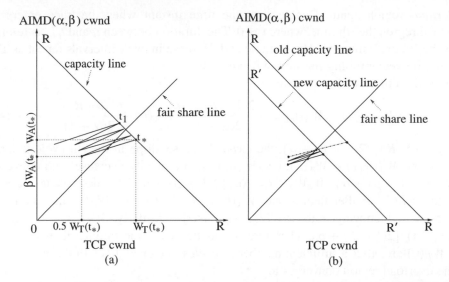

Figure 4.4 *cwnd* trace of one TCP and one AIMD flows: (a) static link capacity; (b) dynamic link capacity [82]. Reproduced by permission of ©2006 IEEE.

(the area under the capacity line), W_T and W_A evolve as follows:

$$W_A(t + \Delta t) = W_A(t) + \alpha \cdot \Delta t, \tag{4.1}$$

$$W_T(t + \Delta t) = W_T(t) + 1 \cdot \Delta t. \tag{4.2}$$

Combining (4.1) and (4.2), we have

$$\frac{W_A(t + \Delta t) - W_A(t)}{W_T(t + \Delta t) - W_T(t)} = \alpha. \tag{4.3}$$

Therefore, the *cwnd* trace in the underload region follows a line segment with slope α.

When $W_T(t) + W_A(t) \geq R$, referred to as the *overload* region (the area above the capacity line), packet losses occur. The AIMD sender and the TCP sender treat packet losses as congestion indications. Assume that the senders of both flows receive the congestion indication and decrease their *cwnd*s simultaneously once they enter the overload region, we have

$$W_A(t) + W_T(t) = R, \tag{4.4}$$

$$W_A(t^+) = \beta W_A(t), \tag{4.5}$$

$$W_T(t^+) = 0.5 W_T(t). \tag{4.6}$$

As indicated in (4.5)–(4.6), the *cwnd* trace has a slope of $(1 - \beta)W_A(t)/[(1 - 0.5)W_T(t)]$ when responding to congestion. Thereafter, the *cwnd* trace is in the *underload* region again. Denote t_i as the time instant when the trace enters the overload region the ith time, where $i > 0$. The duration between t_i and t_{i+1} is referred to as the ith *cycle* interval. The increase and decrease in *cycle* intervals repeat as TCP and AIMD keep probing for unused bandwidth.

Based on (4.1)–(4.6), we have

$$W_A(t_{i+1}) - W_A(t_i) = -\frac{2(1 - \beta) + \alpha}{2(\alpha + 1)} W_A(t_i) + \frac{\alpha \cdot R}{2(\alpha + 1)}. \tag{4.7}$$

If $W_A(t_i) > R\alpha/(2(1 - \beta) + \alpha)$, the *cwnd* decrease slope in the ith *cycle* is larger than α, and $W_A(t_{i+1}) < W_A(t_i)$, i.e., $W_A(t_i)$ monotonically decreases and converges to $R\alpha/(2(1 - \beta) + \alpha)$. If $W_A(t_i) < R\alpha/[2(1 - \beta) + \alpha]$, the decreasing slope in the ith *cycle* is smaller than α, and $W_A(t_{i+1}) > W_A(t_i)$, i.e., $W_A(t_i)$ monotonically increases and converges to $R\alpha/[2(1 - \beta) + \alpha]$. Similarly, $W_T(t_i)$ converges to $2R(1 - \beta)/[2(1 - \beta) + \alpha]$. Therefore, no matter what the initial values of $W_A(0)$ and $W_T(0)$ are, after a sufficient number of *cycles*, the *cwnd* trace of these two flows in the overload region converges to

$$W_A(t_*) = \frac{R \cdot \alpha}{2(1 - \beta) + \alpha}, \tag{4.8}$$

$$W_T(t_*) = \frac{R \cdot 2(1 - \beta)}{2(1 - \beta) + \alpha}, \tag{4.9}$$

where t_* represents the time instant at which the two flows enter the overload region in steady state. From (4.8) and (4.9), in *steady state*, W_A and W_T increase and decrease periodically, oscillating along the line segment with slope α, as shown in Figure 4.4(a). The average *cwnd*s of the AIMD flow and the TCP flow are

$$\overline{W}_A = \frac{(1 + \beta)}{2} W_A(t_*) = \frac{(1 + \beta)\alpha R}{4(1 - \beta) + 2\alpha}, \tag{4.10}$$

$$\overline{W}_T = \frac{(1 + 0.5)}{2} W_T(t_*) = \frac{3(1 - \beta)R}{4(1 - \beta) + 2\alpha}. \tag{4.11}$$

For flows with the same value of round-trip time, the average flow throughput is proportional to the average *cwnd*. That is, to guarantee that the TCP and AIMD flows have a fair share of the link capacity,[4] $\overline{W}_A$ should be equal to $\overline{W}_T$. That is, the midpoint of the *cwnd* trace in steady state should sit on the fair share line. From (4.10) and (4.11), we have the *necessary* and *sufficient* TCP-friendly condition as follows.

Proposition 1 *The necessary and sufficient condition for AIMD(α, β) to be TCP-friendly is*

$$\alpha = \frac{3(1 - \beta)}{1 + \beta}, \quad \text{where } 0 < \alpha < 3 \text{ and } 0 < \beta < 1. \tag{4.12}$$

---

[4]For simplicity, in this section we assume that all flows have the same weight or priority. When they have different weights, as discussed in Subsection 4.5.2, the friendly conditions need to be biased with a scalar offset.

Using the same technique, the TCP-friendly condition can be extended for any two classes of AIMD flows.

Proposition 2 *The necessary and sufficient condition for AIMD(α_1, β_1) and AIMD(α_2, β_2) to be friendly is*

$$\frac{\alpha_1(1 + \beta_1)}{1 - \beta_1} = \frac{\alpha_2(1 + \beta_2)}{1 - \beta_2}, \quad \text{where } \alpha_1, \alpha_2 > 0 \text{ and } 0 < \beta_1, \beta_2 < 1. \tag{4.13}$$

Equations (4.12) and (4.13) are valid as long as the competing flows have the same chance to capture congestion indicators and reduce their *cwnd*s. With random early detection (RED)-capable routers, packets are discarded or marked proportionally when the queue length exceeds a threshold. Therefore, it is reasonable to assume coexisting flows receiving the same share of congestion indicators, unless the *cwnd* of one flow is much larger than that of the other. If so, the flow having the larger *cwnd* has a better chance of receiving congestion indicators and reduces its *cwnd* more frequently; consequently, the *cwnd* trace can attain steady state even faster.

Dynamic link capacity   (4.12) indicates that the TCP-friendly condition is actually independent of the link capacity R. As shown in Figure 4.4(b), when the link capacity changes from R to R', whether $R' > R$ or $R' < R$, the *cwnd* trace converges to a new line segment. When (4.12) is satisfied, the midpoint of the new line segment always sits on the fair share line, which means that the average throughput of the AIMD and TCP flows are the same in steady state. This property is particularly important for congestion control in the Internet, since the end-systems do not have the knowledge of the bottleneck link capacity beforehand.

4.3.2   Multi-class AIMD flows

Different multimedia applications may prefer different (α, β) pairs. Although the TCP-friendly condition is derived by studying the competitive behaviors of two flows, it will be demonstrated that multiple multi-class AIMD flows can coexist in a friendly manner.

Suppose n AIMD flows share a link with capacity R packets per *rtt*. Let the kth AIMD flow have the pair (α_k, β_k), and its *cwnd* at time t be $W_k(t)$. The (α, β) pairs should be pairwise friendly according to (4.13), i.e., $\alpha_k(1 + \beta_k)/(1 - \beta_k)$ is constant for all flows.

In the underload region, their *cwnd* trace evolves as

$$W_k(t + \Delta t) = W_k(t) + \alpha_k \cdot \Delta t. \tag{4.14}$$

When they enter the overload region,

$$\sum_{k=1}^{n} W_k(t) = R, \tag{4.15}$$

$$W_k(t^+) = \beta_k W_k(t). \tag{4.16}$$

The initial value of *cwnd* of the kth flow is $W_k(0)$. Let t_i denote the time instant when the *cwnd* trace enters the overload region for the ith time, and $\Delta t_i = t_{i+1} - t_i$ $(i > 0)$. Δt_0 is the time from the beginning to t_1. From (4.14)–(4.16), we have

$$W_k(t_i) = (W_k(0) + \alpha_k \Delta t_0)\beta_k^{i-1} + \sum_{j=1}^{i-1} \alpha_k \Delta t_j \beta_k^{i-1-j}, \quad \text{for } i > 0. \qquad (4.17)$$

Since $\beta_k < 1$, $\beta_k^{i-1} \to 0$ when i is infinitely large. The contribution of the initial value of *cwnd* damps out exponentially and becomes negligible after several cycles.

In steady state, the random process Δt_i has a constant mean $\mathrm{E}[\Delta t]$.

$$W_k(t_*) = \lim_{i \to \infty} \mathrm{E}[W_k(t_i)] = \mathrm{E}[\Delta t]\frac{\alpha_k}{1 - \beta_k}. \qquad (4.18)$$

The average *cwnd* of the kth flow in steady state is

$$\overline{W}_k = \frac{1 + \beta_k}{2} W_k(t_*) = \frac{\alpha_k(1 + \beta_k)}{2(1 - \beta_k)}\mathrm{E}[\Delta t]. \qquad (4.19)$$

Since all (α, β) pairs satisfy the friendly condition (4.13), $\overline{W}_k$ is constant for all flows. Thus, when their (α, β) pairs satisfy the friendly condition, in the long term, all AIMD flows eventually have an equal share of the link capacity, no matter what the link capacity is, how many AIMD classes exist, or how many flows are from each class.

4.3.3 Variable packet size and *rtt*

TCP is known for its bias against flows with small packets and long *rtt*, and the same is true for the TCP-friendly AIMD (TAIMD). TAIMD flows and TCP flows have the same *packet* rate when they coexist. However, the throughput of an individual TAIMD or TCP flow, in bits per second, is proportional to its average packet size. One possible compensation is to give the small-packet flows a larger weight to have a higher packet rate, which will be discussed in Subsection 4.5.2. For flows with different *rtt*, since an AIMD flow increases its *cwnd* by α packets per *rtt*, the effective increase rate is inversely proportional to its *rtt*. How to achieve fairness for flows with different packet size and *rtt* in highly multiplexed links remains an open issue. In Chapter 6, we will further discuss the fairness issue for flows with different *rtt*s in lightly multiplexed links.

4.3.4 Comparison with other binomial schemes

In steady state, the average value of the *cwnd*s of the competing AIMD flows is independent of the bottleneck capacity. With this property, according to the TCP-friendly condition, the flows in different classes can choose different (α, β) pairs and coexist in a friendly way. For other binomial increase/decrease schemes [87], e.g., inverse increase and additive decrease (IIAD) and square-root increase and square-root decrease (SQRT), although they are attractive to multimedia applications requiring smooth throughput, they cannot achieve TCP-friendliness independent of the link capacity. In steady state, the bandwidth share for an AIMD flow and a coexisting non-AIMD flow always depends on the bottleneck capacity. We demonstrate this using the following example: let an IIAD flow and a TCP flow compete in a link with capacity R packets per *rtt*. Assume that their *rwnds* are always larger than their *cwnds*.

In the *underload* state, the window of the IIAD flow, W, changes as follows:

$$W(t + \Delta t) = W(t) + \frac{\alpha \Delta t}{W(t) \cdot rtt}, \quad \alpha > 0. \tag{4.20}$$

In the *overload* state, W changes in the following way:

$$W(t^+) = W(t) - \beta, \quad 0 < \beta < 1. \tag{4.21}$$

In steady state, IIAD's window oscillates along $(\alpha - \beta)R/(2\beta + R)$ to $(2\beta^2 + \alpha R)/(2\beta + R)$, and TCP's window oscillates from $(R^2 + (2\beta - \alpha)R - 2\beta^2)/(4\beta + 2R)$ to $(R^2 + (2\beta - \alpha)R - 2\beta^2)/(2\beta + R)$. Therefore, the ratio of the average IIAD window and TCP window is not independent of R, i.e., no (α, β) pair can guarantee the IIAD flow to be TCP-friendly with arbitrary link capacity.

In addition, unlike AIMD flows, for IIAD flows to share the link capacity fairly, their (α, β) pair must be the same. To prove this, let two IIAD flows, one with parameter pair (α_1, β_1) and the other with parameter pair (α_2, β_2), compete in a link. In steady state, their average window ratio is

$$\frac{\overline{W_1}}{\overline{W_2}} = \frac{\alpha_2\beta_1^2 - 2\alpha_1\beta_2^2 + 2\alpha_1\beta_2 R - \alpha_1\beta_1\beta_2}{\alpha_1\beta_2^2 - 2\alpha_2\beta_1^2 + 2\alpha_2\beta_1 R - \alpha_2\beta_1\beta_2}. \tag{4.22}$$

Therefore, to achieve the ratio of 1 independent of R, the only solution is $\alpha_1 = \alpha_2$ and $\beta_1 = \beta_2$. In other words, for IIAD flows to be friendly with each other, only one (α, β) pair can be chosen for all flows. Reference [86] has also shown that AIMD is the *only* TCP-friendly binomial control with monotonic convergence to fairness.

4.4 Properties of AIMD

Based on the analysis in Section 4.3, the TCP-friendly property is obvious for AIMD flows satisfying the TCP-friendly condition. Next, we explore two other important properties of AIMD: effectiveness and responsiveness.

4.4.1 AIMD effectiveness

One measure of effectiveness is network *utilization*, i.e., how effectively the AIMD flows can utilize the available bandwidth compared with TCP flows.

Single AIMD flow   Let one AIMD(α, β) flow occupy a link with a fixed capacity. Without queueing delay, the *rtt*, denoted as τ_0, is constant. The link can transmit R packets per τ_0, and the bottleneck buffer size is B packets. For RED-capable routers, B is defined as the mean value of the queue length when packet loss occurs.[5] As shown in Figure 4.5, there are two stages in steady state. In stage I, *cwnd* $\leq R$; there is no queueing delay, and *rtt* $= \tau_0$. In stage II, *cwnd* $> R$ and the queue builds up. For each round,[6] *cwnd* increases by α packets, and so does the queue. Therefore, *rtt* is increased by $\alpha\tau_0/R$ seconds per round. The link utilization is 1 in stage II since the queue is non-empty all the time. When the *cwnd* increases to exceed $R + B$, there are B packets in the queue, and packet loss occurs. Consequently, the *cwnd* is throttled to $\beta(R + B)$. In stage I, the number of packets being transmitted is $B + [R^2 - \beta^2(R + B)^2]/2\alpha$. The time duration in stage I is $\tau_0[R - \beta(R + B)]/\alpha$. Therefore, the link utilization in stage I, η_I, is

$$\eta_I = \begin{cases} \dfrac{R^2 - \beta^2(R + B)^2 + 2\alpha B}{2R^2 - 2\beta R(R + B)}, & \text{when } 0 \leq B \leq B', \\ 1, & \text{otherwise,} \end{cases} \quad (4.23)$$

where

$$B' = \left(\sqrt{\frac{\alpha}{2\beta^2} + \left(\frac{1}{\beta} - 1\right)R} - \sqrt{\frac{\alpha}{2\beta^2}} \right)^2.$$

Equation (4.23) indicates that TCP-friendly AIMD flows with a larger value of β only require a relatively small buffer to fully utilize the link. On the other hand, although a large buffer can improve the link utilization, it also introduces more delay, which is undesirable for delay-sensitive applications. Thus, it is preferable to have a large value of β and moderate buffer size.

Without a buffer, the link utilization for an AIMD flow is

$$\eta = \frac{1 + \beta}{2} \quad \text{where } 0 < \beta < 1. \quad (4.24)$$

Equation (4.24) indicates that the effective link utilization of an AIMD flow is proportional to $1 + \beta$.

Multiple AIMD flows   When there are n AIMD(α, β) flows sharing a link, the link utilization becomes much more complicated to analyze since it also depends on whether the congestion indicators are synchronized or not.

The worst case scenario is that all flows simultaneously increase or decrease their windows. The link utilization in such cases is exactly the same as when only one AIMD($n\alpha, \beta$) flow traverses the link. If network buffering is neglected, the link utilization in the worst case, η_{worst}, is the same as (4.24).

---

[5]Detailed discussions on the interaction between AIMD and RED will be given in Chapter 5.

[6]Here, a round is defined as the time interval to successfully transmit and receive a whole *cwnd* of packets and acknowledgments.

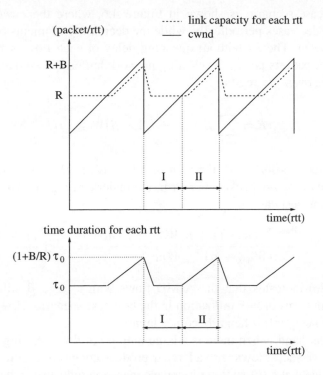

Figure 4.5 *cwnd* vs. throughput for AIMD flows [82]. Reproduced by permission of
©2006 IEEE.

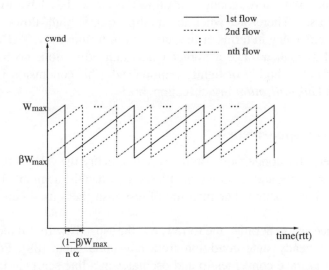

Figure 4.6 *n* AIMD(α, β) flows, best scenario [82]. Reproduced by permission of
©2006 IEEE.

The best case scenario is shown in Figure 4.6, where the *cwnd* of each flow increases and decreases periodically, while the decreasing moments of each flow are evenly distributed. The *rtt* without queueing delay of each flow is τ_0, and the link can transmit R packets per τ_0. Neglecting network buffering, we have the following equation in the *overload* region:

$$R = \sum_{i=1}^{n}(\beta W_{max} + i(1-\beta)W_{max}/n). \tag{4.25}$$

After a flow reduces its *cwnd*, the sum of all *cwnd*s is $R - (1-\beta)W_{max}$, and the sum will increase by $n\alpha$ packets per *rtt* in the underload region. Therefore, the link utilization in the best case is

$$\eta_{best} = \frac{\sum_{i=0}^{(1-\beta)W_{max}/n\alpha}(R-(1-\beta)W_{max}+i\cdot n\alpha)}{R(W_{max}-\beta W_{max})/n\alpha} = 1 - \left(\frac{(1+\beta)n}{1-\beta}+1\right)^{-1}. \tag{4.26}$$

Equation (4.26) indicates that the AIMD flows with a large β still exhibit better performance in terms of link utilization in the best case scenario. However, the effect of β becomes negligible when n becomes large.

On the other hand, when there is a large number of flows sharing a link, the link is frequently overshot. Flows with a large α produce congestion more frequently. For instance, assume that k (of n) flows receive congestion indicators when overshooting the link capacity R. The time duration to the next congestion moment is $R(1-\beta k/n)/(n\alpha)$. The congestion frequency is approximately proportional to α. Frequent congestion reduces the throughput of all flows, because they have to recover from more packet losses. Therefore, when the multiplexing is high, flows with a smaller α have higher efficiency in terms of link utilization. In summary, TAIMD flows with a large β and a small α have a higher bandwidth utilization, no matter whether the bottleneck link is highly or lightly multiplexed. This conclusion is validated by simulations in *Link utilization* in Subsection 4.6.1.

4.4.2   AIMD responsiveness

Another property for congestion control is how quickly the AIMD-based protocols respond to the network and session dynamics, e.g., when the available bandwidth and the number of flows change. The measure of responsiveness used here is *convergent rate*.

Convergence is generally measured by the speed with which the system approaches the steady state condition from any initial state [88]. For two AIMD flows, their *cwnd* trace converges to and oscillates in a line segment in steady state. If these two AIMD flows have the same (α, β) pair, this line segment coincides with their fair share line. Otherwise, if their AIMD parameter pairs satisfy the friendly condition, the converged line segment intersects the fair share line at its midpoint.

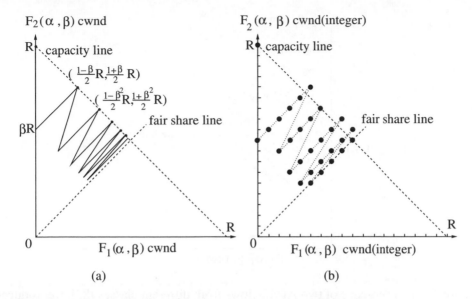

Figure 4.7 Convergence of two AIMD flows from the same class: (a) *cwnd* is a continuous variable; (b) *cwnd* is an integer [82]. Reproduced by permission of ©2006 IEEE.

Single AIMD class   We first consider two flows from the same AIMD(α, β) class, F_1 and F_2, sharing a link with capacity R. F_2 takes the whole capacity before F_1 shares the link. Initially, their *cwnd* trace is at the point $(0, R)$, as shown in Figure 4.7(a). After n decrease and increase *cycles*, their *cwnd* trace is at the point $((1 - \beta^n)R/2, (1 + \beta^n)R/2)$. Since both flows have the same (α, β) pair, each *cycle* takes the same amount of time, which equals $(1 - \beta)R/(2\alpha)$ *rtts*. The convergent speed, denoted as c, is the speed in gaining bandwidth for F_1; it equals the amount of bandwidth gained by F_1 taken over the time interval measured in *rtts*:

$$c(\alpha, \beta) = \frac{(1 - \beta^n)R/2}{n(1 - \beta)R/2\alpha} = \frac{\alpha(1 - \beta^n)}{n(1 - \beta)}. \tag{4.27}$$

From (4.27), the larger the α or β, the faster the convergence. However, if the (α, β) pair is under the constraint of the TCP-friendly condition in Proposition 1, we cannot enlarge both α and β to attain a higher convergent speed. By substituting (4.12) into (4.27), the convergent speed of the TAIMD flow is

$$c(\beta) = \frac{3(1 - \beta)}{(1 + \beta)} \frac{(1 - \beta^n)}{n(1 - \beta)} = \frac{3(1 - \beta^n)}{n(1 + \beta)}. \tag{4.28}$$

Therefore, the speed for F_1 in gaining A portion of the total bandwidth is

$$c(\beta) = \frac{6A}{(1 + \beta) \log_\beta (1 - 2A)}, \tag{4.29}$$

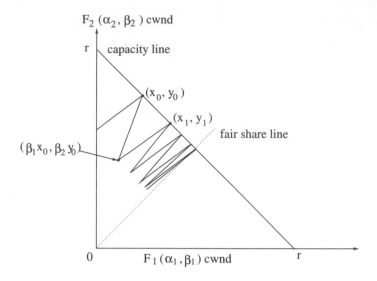

Figure 4.8 Convergence of two AIMD flows from different classes [82]. Reproduced by permission of ©2006 IEEE.

where $0 < A < 0.5$. From (4.28) and (4.29), c, in packets per rtt^2, increases when β decreases. Therefore, when the same-class TAIMD flows compete, the convergent speed is proportional to $\log(1/\beta)/(1 + \beta)$.

If $cwnd$ is a $continuous$ variable, it will take infinite time for the $cwnd$ trace to reach steady state (when $A \to 0.5$, $\log_\beta(1 - 2A) \to \infty$). However, in reality, $cwnd$ is a non-negative integer because of packetization, so within finite rounds of increase and decrease $cycles$, the $cwnd$ trace attains steady state, as shown in Figure 4.7(b).

Multiple AIMD classes   Let two flows from different AIMD classes, $F_1(\alpha_1, \beta_1)$ and $F_2(\alpha_2, \beta_2)$, share the link, as shown in Figure 4.8. Initially, their $cwnd$ trace hits the capacity line at the point (x_0, y_0), where $x_0 + y_0 = r$. After one decrease and increase $cycle$, the $cwnd$ trace is at the point (x_1, y_1), where x_1 and y_1 are given by

$$x_1 = \frac{R - \beta_2 y_0 + (\alpha_2/\alpha_1)\beta_1 x_0}{1 + (\alpha_2/\alpha_1)}, \tag{4.30}$$

$$y_1 = R - x_1. \tag{4.31}$$

The bandwidth gained by F_1 is $(x_1 - x_0)$, and the number of rtts it takes to achieve this gain is $(x_1 - \beta_1 x_0)/\alpha_1$. Then, the convergent speed in this $cycle$ is

$$c(\alpha_1, \beta_1; \alpha_2, \beta_2) = \frac{(R - \beta_2 y_0 - x_0)\alpha_1 + (\beta_1 - 1)\alpha_2 x_0}{R - \beta_2 y_0 - \beta_1 x_0}. \tag{4.32}$$

The partial derivatives of c with respect to α_1 and β_1 are both positive, while the partial derivatives of c over α_2 and β_2 are both negative. This means that for F_1 to gain the bandwidth faster, its own α and β values should be larger, while the α and β values for the competing flow should be smaller. Furthermore, if the (α, β) of F_1 are under the constraint of the TCP-friendly condition, the convergent speed becomes

$$c(\alpha_1, \alpha_2, \beta_2) = \frac{R - \beta_2 y_0 - x_0}{R - \beta_2 y_0 + x_0} - \frac{2\alpha_2 x_0}{(R - \beta_2 y_0 + x_0)\alpha_1 + 3(4 - \beta_2 y_0 - x_0)}. \quad (4.33)$$

The partial derivative of $c(\alpha_1, \alpha_2, \beta_2)$ with respect to α_1 is always positive, which means that c is larger if α_1 is larger. The analysis of the speed in gaining bandwidth for the new arriving flow considers the situation that both flows throttle their *cwnds* simultaneously when they overshoot the available bandwidth. In reality, when the new flow has a much smaller *cwnd* than that of the old flow, it is likely that only the old flow suffers the packet loss and shrinks its *cwnd* [89]. In that case, the speed in gaining bandwidth for the new arrival is even faster.

It is observed that, for a new TAIMD flow to get its fair share bandwidth faster, its α should be larger and β should be smaller. Simulation results in *Convergent speed* in Subsection 4.6.1 confirm this observation. The above analysis focuses on the congestion avoidance and exponential backoff phases, and it is helpful in determining how fast AIMD flows respond to the change of available bandwidth in the network.

4.4.3 Practical implications

The achievable *cwnd* of an AIMD flow is determined by the available link capacity and *rtt*. When *cwnd* is very small due to a limitation in link capacity, the AIMD flow with a larger β has less throughput than the competing TCP flow, even if its (α, β) pair satisfies the friendly condition. In the following we discuss the practical implications that affect the TCP-friendliness of AIMD flows.

The TCP-friendly condition is derived with the assumption that *cwnd* is a continuous variable. In reality, because of packetization, the effective *cwnd* is always rounded to the maximum integer no larger than the algorithmic *cwnd*. When *cwnd* is small, the effect of this rounding on the throughput is non-trivial. If *cwnd* is smaller than $1/(1 - \beta)$ when congestion occurs, it has to be reduced at least by one packet in response to the congestion indicator, although theoretically $(1 - \beta)$ times *cwnd* is smaller than 1. Therefore, the effective decrease ratio is $1 - 1/cwnd$, which is smaller than β. When the available link capacity to each flow is so small that the average *cwnd* of each flow is smaller than $1/(1 - \beta)$, the throughput of AIMD flows is much less than the theoretical result. It follows that the larger is β, the larger will be $1/(1 - \beta)$, and thus there is more negative effect on the throughput. Therefore, the AIMD flow with a large β and a small α has less throughput than that of TCP flows.

In addition, for TAIMD flows, if β is larger than 0.5, α should be smaller than 1. If there is no congestion, it takes $1/\alpha$ *rtts* to increase the effective *cwnd* by 1. During this time, if a single packet loss occurs, the effective *cwnd* has no chance to increase before it actually decreases, so the effective increase rate is smaller than α. If the *cwnd* is small when congestion occurs, the difference between α and the effective increase rate is non-negligible. Therefore, AIMD flows with a large β and a small α may have less throughput than the competing TCP flows, especially when *cwnd* is small.

4.4.4   An enhanced AIMD algorithm – DTAIMD

To mitigate the practical implications, consider the DTAIMD algorithm. Since the effective decrease ratio, β', is $1 - 1/cwnd$ when $cwnd < 1/(1 - \beta)$, α should be adjusted based on β', instead of β, according to the TCP-friendly condition. When $cwnd \geq 1/(1 - \beta)$, the increase ratio is the preset value α, which equals $3(1 - \beta)/(1 + \beta)$; when $cwnd < 1/(1 - \beta)$, the increase ratio should be $3(1 - \beta')/(1 + \beta')$, or $3/(2cwnd - 1)$, i.e., α should be dynamically adjusted according to the current *cwnd* to ensure the practical TCP-friendliness.

The core pseudo code for the DTAIMD algorithm based on the above analysis to enhance the TCP-friendliness is listed in Algorithm 1.

Algorithm 1 DTAIMD algorithm [81]. Reproduced by permission of ©2005 IEEE.

1: **if** $cwnd \geq 1/(1 - \beta)$ **then**
2:     $\alpha = 3(1 - \beta)/(1 + \beta)$;
3: **else**
4:     **if** $cwnd = 1$ **then**
5:         $\alpha = 1$;
6:     **else**
7:         $\alpha = 3/(2\,cwnd - 1)$;
8:     **end if**
9: **end if**

If *cwnd* is large enough ($cwnd > 1/(1 - \beta)$), the regular TCP-friendly condition applies (Line 2). Otherwise, *cwnd* is adjusted according to the effective $\beta' = 1 - 1/cwnd$ (Line 7), unless *cwnd* is minimized already (Line 4). The algorithm is generic and adjusts α adaptively, and it still conserves the desired AIMD properties. We will evaluate its performance and compare it with the ordinary AIMD and TCP in Section 4.6.

To give an insight into the efficiency in applying the AIMD protocol in handling Internet services, we present in the next section a case study of the AIMD protocol supporting Internet-based multimedia playback applications with service differentiation.

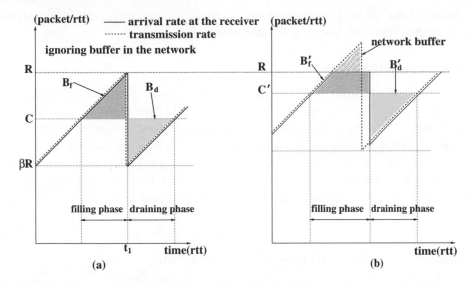

Figure 4.9 Filling and draining phases: (a) without network buffer; (b) with network buffer [82]. Reproduced by permission of ©2006 IEEE.

4.5   Case Study: Multimedia Playback Applications with Service Differentiation

4.5.1   Multimedia playback applications

Web-based multimedia playback applications are becoming popular, and are advocated as the future of home entertainment. The multimedia stream length varies over a wide range, from clips of a few seconds to hour-length movies. Users normally desire the best affordable quality in terms of higher application throughput, minimal start-up latency and fewer interruptions during playback. Let the DTAIMD-based protocol be the mechanism for transporting multimedia playback applications. Appropriate (α, β) pairs should be chosen to achieve the TCP-friendliness and better quality provisioning for these applications.

To simplify the analysis, let the playback application consume a stream at constant rate C [90]. This assumption is reasonable as a starting point, and many schemes [91] have been proposed to smooth streams for easier resource reservation. These schemes can also be used in our context.

As shown in Figure 4.9(a), if the packet arrival rate at the receiver (equal to the transmission rate when network buffering is ignored) is above C, the excess data fills the receiver buffer. This time period is referred to as the *filling* phase. At time t_1, packet loss occurs and the transmission rate is reduced to below C; some data must be drained from the receiver buffer until the transmission rate reaches C again, and this time interval is referred to as the *draining* phase.

Low multiplexing scenario

Consider a static situation in which only one AIMD flow traverses a link of capacity R packets per *rtt*. The transmission rate of the AIMD flow increases and decreases periodically. When the transmission rate reaches R, *cwnd* decreases to βR. If the amount of data stored during the filling phase, B_f, is less than the amount of data drawn from the receiver buffer during the draining phase, B_d, the receiver will starve and the playback will freeze for a while, which is undesirable for playback applications. To guarantee $B_f \geq B_d$, the consumption rate C should be bounded as follows:

$$C \leq (1 + \beta)R/2. \tag{4.34}$$

This is because, as shown in Figure 4.9(a), we can obtain

$$B_f/B_d = [(R - C)/(C - \beta R)]^2. \tag{4.35}$$

Equation (4.34) indicates that the maximum consumption rate, C_{max}, of an AIMD(α, β) flow depends on the value of β. The minimal receiver buffer size required (maximum queue length) is $[R^2(1 - \beta^2) + 6R(1 - \beta)]/24$. Therefore, with a larger β, the receiver buffer size can be reduced significantly, and so the maximum delay in the receiver buffer reduces.[7] For instance, C of an AIMD$(1, 0.5)$ flow is 25% less than that of an AIMD$(0.2, 0.875)$ flow. A higher permissible consumption rate allows potentially higher application throughput and better audio or video quality. In other words, the AIMD protocol with a larger β can provide better QoS. On the other hand, since a network buffer has the effect of flattening the arrival rate to the receiver, C_{max} can be enlarged with the network buffer size, and the advantage of large β may become less significant, as shown in Figure 4.9(b).

High multiplexing scenario

When many AIMD flows with different (α, β) pairs compete for a highly multiplexed bottleneck, C_{max} of a flow cannot be determined only by its own β as in (4.34). If the (α, β) pairs of coexisting flows satisfy the friendly condition, each flow has the same average throughput and C_{max}. However, in a highly multiplexed dynamic network, AIMD flows with different (α, β) pairs have different response patterns to transient changes of the network resources. AIMD flows with a small value of β and a large value of α are very sensitive to bandwidth variation, and their instantaneous throughput changes quickly. (It has been shown that the convergent speed of the

---

[7]In this section, we focus on the delay in the receiver buffer, i.e., the delay between the instant the packet arrives at the transport layer receiver and the instant the packet is being delivered to the application receiver. The end-to-end delay, which measures the time between a packet being sent and it being received by the transport layer, will be discussed in Chapter 6.

TAIMD flow is inversely related to β in Subsection 4.4.2.) Therefore, during the playback, they are more likely to be interrupted. The frequency of interruption, or buffer underrun, is an important QoS index for multimedia playback applications, and the AIMD protocol with a larger value of β is desirable for these applications, as shown in Subsection 4.6.2.

4.5.2   Service differentiation

Different multimedia applications can have different service requirements. To provide heterogeneous services, we can categorize the traffic into several classes and allocate to each class a different priority of occupying the network resources (e.g., bandwidth). If a family of AIMD parameters can be used to provide such differentiation, no major upgrades to the core network are required. It is a scalable and economical approach to offering service differentiation in the Internet.

From the analysis in Section 4.3, the *cwnd* trace of coexisting AIMD flows converges to a steady state. The average of the *cwnd*s of two flows in steady state depends on their (α, β) pairs and is independent of the link capacity. For two traffic classes, let class 2 have a higher priority than class 1: class 1 flows are weighted as 1 and class 2 flows are weighted as an integer k (> 1). When they compete in the same bottleneck link, the throughput of class 2 AIMD(α_2, β_2) flows can be k times that of class 1 AIMD(α_1, β_1) flows if their average *cwnd* ratio is k.

Let $t_i^{(1)}$ and $t_i^{(2)}$ denote the ith time that the AIMD(α_1, β_1) and AIMD(α_2, β_2) flows are in the *overload* region, respectively. $\Delta t_i^{(\cdot)}$ is the time duration between $t_{i-1}^{(\cdot)}$ and $t_i^{(\cdot)}$. Their *cwnd*s are given by

$$W_1(t_n^{(1)}) = W_1(0)\beta_1^{n-1} + \alpha_1 \sum_{i=1}^{n-1} \Delta t_i^{(1)} \beta_1^{n-i}, \tag{4.36}$$

$$W_2(t_n^{(2)}) = W_2(0)\beta_2^{n-1} + \alpha_2 \sum_{i=1}^{n-1} \Delta t_i^{(2)} \beta_2^{n-i}. \tag{4.37}$$

From the above equations, we can get the average *cwnd* of each of the two flows in steady state:

$$E[W_1] = \frac{\alpha_1 E[\Delta t^{(1)}](1 + \beta_1)}{2(1 - \beta_1)}, \tag{4.38}$$

$$E[W_2] = \frac{\alpha_2 E[\Delta t^{(2)}](1 + \beta_2)}{2(1 - \beta_2)}. \tag{4.39}$$

If the average sending rate of the AIMD(α_2, β_2) flow is k times that of the AIMD(α_1, β_1) flow, with RED-capable routers, the packet loss rate of AIMD(α_2, β_2) is also k times that of AIMD(α_1, β_1). In other words, the average duration, $E[\Delta t^{(1)}]$, of increase and decrease cycle of AIMD(α_1, β_1) is $1/k$ of

$E[\Delta t^{(2)}]$. Therefore, according to (4.38) and (4.39), to ensure $E[W_2] = kE[W_1]$, their (α, β) pairs should satisfy

$$\frac{\alpha_2}{\alpha_1} = \frac{k^2(1 + \beta_1)(1 - \beta_2)}{(1 + \beta_2)(1 - \beta_1)}. \tag{4.40}$$

For instance, let the data traffic be class 1 with $\alpha_1 = 1$ and $\beta_1 = 0.5$, and the multimedia traffic be class 2 with $\beta_2 = 0.875$. From (4.40), α_2 takes the values 0.8, 1.8, 3.2 for $k = 2, 3, 4$, respectively. With the (α, β) values of (3.2, 0.875), a multimedia flow can have a four times higher packet transmission rate than the coexisting TCP flows. In *Service differentiation* in Subsection 4.6.2, we will illustrate this property via simulation.

Implementation concerns   The values of the (α, β) pair can be flexibly chosen according to the application's QoS requirements. To prevent abusive users or misbehaving applications from monopolizing network resources, it is suggested that an application can choose only one parameter, and it informs the transport layer of its desired parameter value. The operating system and protocol stack calculate the other one according to the TCP-friendly condition and the service differentiation weight. Obviously, this still cannot fully prevent malicious users from abusing the network by hacking the transport layer protocol or operating systems. The stability of the Internet relies on voluntary cooperation from all end users. How to protect the network from malicious users is beyond the scope of this book, and the DTAIMD protocol does not introduce any additional risks compared with the current TCP/IP stack.

4.6   Performance Evaluation

Computer networks are very dynamic and complex systems, and it is very difficult if not impossible to analyze the network performance considering all network dynamics and protocol details. To validate the analytical results and evaluate network algorithms and protocols, extensive experiments with different network scenarios are needed. However, real networking experiments in a controlled and repeatable environment are very costly and time-consuming, while network simulation is low-cost to assemble and can be easily configured to examine a wide range of scenarios in a relatively short amount of time. Therefore, network simulations are widely used for networking research. NS-2 is the most popular open-source network simulator. In this section, we present a performance study made by extensive simulations using NS-2.

To validate the TCP-friendly condition derived in Section 4.3, we examine the bandwidth utilization and convergent speed analyzed in Section 4.4, and compare the QoS provided by DTAIMD-based algorithms with Selective Acknowledgement (SACK)-based TCP for the multimedia applications discussed in Section 4.5.

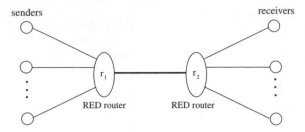

Figure 4.10 Simulation topology.

4.6.1 Performance of AIMD algorithms

The logical simulation topology shown in Figure 4.10 is the widely used shared bottleneck topology. In the previous analysis sections, the slow start phase of the TCP and AIMD protocols is ignored. In the simulations, the full congestion control algorithms for TCP and AIMD are implemented. The simulations also capture all delay components including random queueing delay in the network. The simulations use the following parameters and settings unless otherwise explicitly stated. The routers adjacent to the bottlencck link are RED-capable and randomly discard incoming packets when the queue length exceeds the predefined threshold. The link propagation delay is 20 ms. The number of competing flows sharing the bottleneck link ranges from 2 to 128, which covers the low multiplexing and high multiplexing (congestion) scenarios. To avoid the *phase effect*,[8] the *rtt* of each flow is made slightly different. The packet size is 1000 bytes.

Flow fairness

We first evaluate the fairness among TCP, TAIMD, and DTAIMD flows in terms of *normalized throughput*, defined as the ratio of the flow throughput to their fair share. The normalized throughput should be 1 if the flow throughput equals the fair share. Figure 4.11(a) shows that TCP flows have a higher average throughput than those of TAIMD(0.2, 0.875) flows when they compete for the bandwidth of the same link, especially when the number of flows, n, sharing the same link is large and the average *cwnd* is small. This observation reveals the practical fairness problem between TCP and the ordinary TAIMD as discussed in Subsection 4.4.3. Figure 4.11(b) shows that the normalized throughputs of TCP and DTAIMD(0.2, 0.875) are close to each other, which implies better fairness than those shown in Figure 4.11(a). This demonstrates that the DTAIMD algorithm works properly and does improve the fairness between the TCP and AIMD flows.

---

[8]Window control protocols are cyclic with period equal to the connection round-trip time. Control theory suggests that this periodicity can resonate (i.e., have a strong, nonlinear interaction) with deterministic control algorithms in network gateways, which results in the phase effect [92, 93].

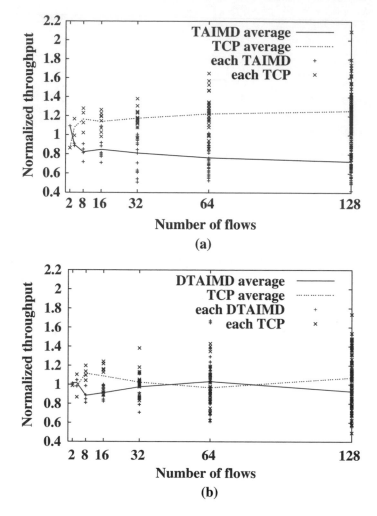

Figure 4.11 Normalized throughput (RED, 15 Mbps): (a) TAIMD(0.2, 0.875) vs. TCP; (b) DTAIMD(0.2, 0.875) vs. TCP [82]. Reproduced by permission of ©2006 IEEE.

To further clarify the fairness performance, we plot the *fairness index* in Figure 4.12 for TCP, TAIMD(0.2, 0.875), and DTAIMD(0.2, 0.875) flows. The fairness index is defined as

$$\left(\sum_{i=1}^{n} f_i\right)^2 \Big/ \left(n \sum_{i=1}^{n} f_i^2\right),$$

where f_i is the throughput of the ith flow [94]. The fairness index ranges from $1/n$ to 1, and reaches 1 when all flows have the same throughput. Figure 4.12 shows

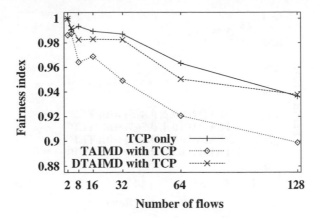

Figure 4.12 Fairness index (RED, 15 Mbps) [82]. Reproduced by permission of ©2006 IEEE.

that the fairness index of DTAIMD(0.2, 0.875) flows with TCP flows is much better than that of TAIMD(0.2, 0.875) flows with TCP flows. It also shows that the fairness index of DTAIMD flows mixed with TCP flows is almost the same as that of TCP flows alone, which indicates that DTAIMD is truly TCP-friendly over a wide range of network and session dynamics. In the simulations, when there are more than 32 flows sharing a link of 15 Mbps, the average *cwnd* is smaller than three packets, and the congestion is mostly indicated by timeouts. Simulation results show that the throughputs of DTAIMD flows are still close to those of TCP flows, which shows that the derived TCP-friendly condition is valid even with frequent timeouts. Simulations with other TCP-friendly parameter pairs exhibit similar results, as shown in Figure 4.13.

Link utilization

To examine the link utilization derived in Section 4.4.1, four AIMD(α, β) pairs, (0.2, 0.875), (0.4, 0.765), (0.6, 0.667) and (1, 0.5), are chosen according to the TCP-friendly condition. Each simulation runs for 30 seconds, and the results are collected after five seconds from the beginning to avoid the skew introduced by the warming-up effect.

Single TAIMD flow   In this set of simulations, one TAIMD flow traverses a 3 Mbps link with a Drop-Tail queue.[9] The *rtt* is around 40 ms. The buffer size is set to 2, 5, 10 or 20 packets. Two is the minimum buffer size used in the simulation since the TAIMD sender sometimes sends two packets back-to-back, e.g., when it receives an *ack* in slow start.

---

[9] A Drop-Tail queue is a FIFO queue, and all incoming packets are dropped when the buffer is full.

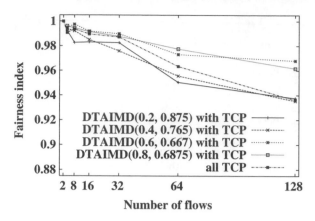

Figure 4.13 Fairness index, TCP with DTAIMD (RED, 15 Mbps) [82]. Reproduced by permission of ©2006 IEEE.

Figure 4.14 shows the results of four different TAIMD(α, β) pairs along the x-axis, and the bottleneck utilizations in the y-axis. It is observed that the analytical results for $B = 0$ closely approximate those of the simulation for $B = 2$. In addition, the figure shows that link utilization is proportional to β, and a TAIMD flow with a larger β needs less buffer to achieve a higher utilization.

Multiple TAIMD flows   Now let two or four TAIMD flows share the link. The buffer size is 2, 5, 10 or 20 packets for the two or four flows cases, respectively. Figures 4.15(a) and 4.15(b) compare the link utilization with Drop-Tail and RED queues when there are four flows. When these flows share the link with the Drop-Tail queue, the link utilization with the minimum buffer size is very close to the analytical results of the worst case scenario, since all flows are synchronized when the buffer overflows. However, the link utilization with the RED queue is very close to the analytical results of the best case scenario. This shows that the RED-capable router can improve the link utilization by letting the competing flows react to congestion evenly in different phases.

We further increase the number of flows in the link to 8 and 16. The minimum buffer size is five packets since it is very likely that more flows may simultaneously transmit two packets back-to-back and need more buffer to accommodate the traffic burst. The link utilizations with both Drop-Tail queue and RED queue are similar to the best case analytical results, as shown in Figures 4.16(a) and (b). This phenomenon reveals that when the link is highly multiplexed with a larger buffer size, the flows are not strictly synchronized as in the worst case scenario, and higher link utilization can be achieved.

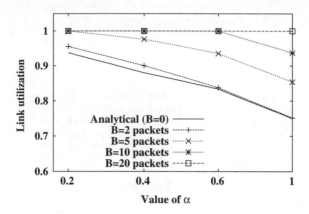

Figure 4.14 Link utilization for one TAIMD(α, β) flow [82]. Reproduced by permission of ©2006 IEEE.

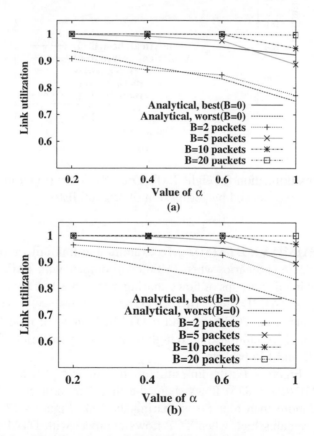

Figure 4.15 Link utilization for four TAIMD(α, β) flows: (a) Drop-Tail queue; (b) RED queue [82]. Reproduced by permission of ©2006 IEEE.

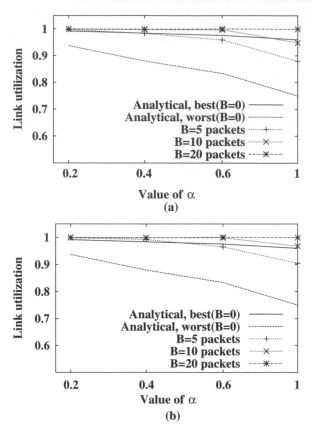

Figure 4.16 Link utilization for eight TAIMD(α, β) flows: (a) Drop-Tail queue; (b) RED queue [82]. Reproduced by permission of ©2006 IEEE.

TCP vs. TAIMD vs. DTAIMD   To examine the link utilization for both low and high multiplexing scenarios when TCP flows compete with TAIMD or DTAIMD flows, the following five groups of flows sharing a 15 Mbps link are simulated: (i) $2n$ TCP flows; (ii) $2n$ TAIMD flows; (iii) $2n$ DTAIMD flows; (iv) n TCP flows and n TAIMD flows; and (v) n TCP flows and n DTAIMD flows. n ranges from 1 to 64, which covers from low multiplexing to high multiplexing scenarios. The RED queue thresholds are 25 and 125 packets.

As shown in Figure 4.17, the link utilization is quite high when more than four TCP or DTAIMD(0.2, 0.875) flows share the link. The utilization is around 99% when there are more than four flows sharing the link. Figure 4.17 shows that the link utilization remains high when TCP flows compete with DTAIMD(0.2, 0.875) flows. This fact indicates that DTAIMD(α, β) is as efficient as TCP in terms of link utilization. Figures 4.17(a) and (b) show that the link utilization is even better for TAIMD(0.2, 0.875) flows only or for TAIMD(0.2, 0.875) flows mixed with

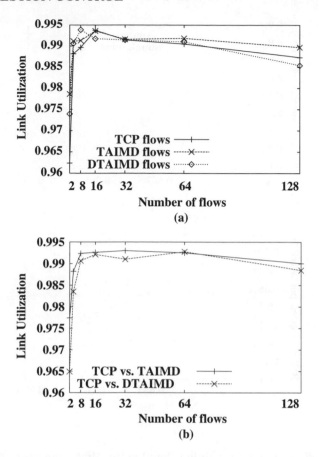

Figure 4.17   Link utilization (RED, 15 Mbps): (a) homogeneous flows;
(b) heterogeneous flows [82]. Reproduced by permission of ©2006 IEEE.

TCP flows. This is because TAIMD(0.2, 0.875) has a large value of β and a small value of α, which is more efficient than TCP in link utilization, whether the link multiplexing is high or low, as discussed in Subsection 4.4.1.

Convergent speed

Figure 4.18(a) shows the results for the case with one AIMD class. The solid line is the analytical result for the convergent speed obtained in (4.29) and the dashed line is the simulation result. Let one DTAIMD flow occupy the whole link at the beginning and the other one share the link later. The speed in gaining bandwidth of the second one is measured. In this set of simulations, the slow start phase is frozen, so both flows increase and decrease their *cwnd*s following the AIMD mechanism only. We set the link buffer to the minimum value; therefore, both flows receive the

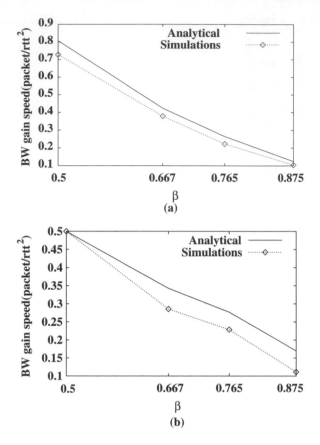

Figure 4.18 F_1's speed in gaining bandwidth (link capacity 20 packets/*rtt*):
(a) 35% bandwidth, one AIMD class; (b) first two cycles, two AIMD classes [82].
Reproduced by permission of ©2006 IEEE.

congestion indicators once the sum of their sending rates exceeds the link capacity.
The β for DTAIMD flows enumerated in the x-axis varies from 0.5 to 0.875, and
α is set according to the TCP-friendly condition. Figure 4.18(a) shows how fast
a new DTAIMD flow gains 35% of the total link capacity from an existing same-
class DTAIMD flow. For TCP, or TAIMD(1, 0.5), it is fairly quick to grab resources
from competing flows. The convergence speed c drops quickly when β grows, and
it confirms that c is inversely proportional to β, as derived in (4.28). In the analysis,
we assume that the *cwnds* are continuous variables while in reality the *cwnds* are
rounded when decreasing. Therefore, the simulation results are slightly lower than
the c we derived.

Similar simulations have been done for multi-class AIMD flows. Figure 4.18(b)
shows the speed in gaining bandwidth for a new DTAIMD flow competing with

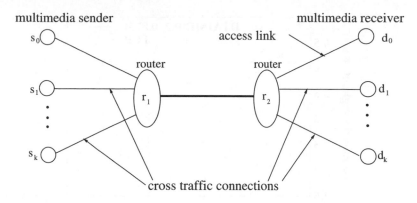

multimedia sender   multimedia receiver

s_0 access link d_0

router   router

s_1 d_1

r_1 r_2

s_k d_k

cross traffic connections

Figure 4.19   Topology of the simulation with cross traffic [82]. Reproduced by permission of ©2006 IEEE.

an existing TCP flow for the same bottleneck. The speed for the DTAIMD flow in gaining bandwidth in the first two cycles is plotted in Figure 4.18(b), since TCP gains most of its share in the first two cycles and its bandwidth gaining speed approaches to zero after that. Since the DTAIMD flow rounds off its *cwnd* when decreasing, the simulation results are slightly lower than the analytical ones, especially when β_1 is much larger than the decrease ratio of the coexisting TCP. Nevertheless, it reveals a similar trend that the larger the β, the slower the convergent speed a DTAIMD flow has, as it becomes less aggressive to grab resources from others.

4.6.2   QoS for multimedia playback applications

Figure 4.19 shows the simulation configuration for multimedia applications: playback streams are stored as files in the server s_0 which is connected to router r_1. The destination receiver d_0 is connected to the network through router r_2. The target multimedia application shares the network resources with other background traffic. In the current Internet, there are two kinds of dominant background traffic, *elephants* and *mice*. The *elephants*, such as long-lived ftp connections, share the link with the target flow throughout the simulation. The *mice*, such as Web transactions, usually have a very short lifetime. We consider both kinds of background traffic. *Mice* start randomly during the simulation and their lifetime follows an exponential distribution with a mean of five seconds.

Access bottleneck

Let the bottleneck of the target multimedia connection be the access line. Examples include a dial-up modem connecting a home PC to an ISP, or a wireless link connecting a PDA to the Internet. The access link bandwidth, $d_0 r_2$, is 56 Kbps. There are ten cross traffic connections, which are *elephants*, sharing link $r_1 r_2$, which has

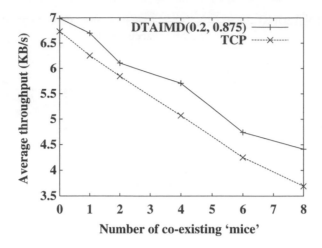

Figure 4.20   Average throughput for the multimedia flow [82]. Reproduced by permission of ©2006 IEEE.

a capacity of 10 Mbps. In addition to the multimedia flow, there are n short-lived *mice* sharing the access link d_0r_2 and the backbone link r_1r_2 with the target flow. n varies from zero to ten. The packet size is 100 bytes for multimedia traffic and 1000 bytes for other traffic. DTAIMD(0.2, 0.875) and TCP are chosen to transport the multimedia flow separately. Figure 4.20 compares their average throughputs. It shows that the average throughput of the DTAIMD(0.2, 0.875) flow is consistently higher than that of the TCP flow. Therefore, DTAIMD(0.2, 0.875) can have a higher playback rate than TCP for multimedia applications, with changing background flows. Figure 4.20 also indicates that the number of simultaneous connections in the access link should be minimized to achieve a higher playback rate for multimedia applications.

Backbone bottleneck

Now let the connection bottleneck be the highly multiplexed backbone link. The multimedia receiver buffers data for several seconds before playback. (Several algorithms have been introduced in [95] to set the playback start-up delay.) Ideally, the consumption rate equals the mean transmission rate. When the buffer becomes empty, the playback is interrupted and freezes for a while. In the simulation, one DTAIMD(0.2, 0.875) flow competes with 10 TCP *elephants* and 20 TCP *mice*, sharing the bottleneck link r_1r_2. The simulation lasts 600 seconds. The multimedia receiver initially stores the data for 3 seconds before it starts to playback at a constant rate, which equals the mean flow throughput. The DTAIMD(0.2, 0.875) flow and the TCP *elephants* are set to support multimedia playback applications separately. To compare the application performance transported by DTAIMD(0.2, 0.875) and TCP,

Table 4.2 QoS for multimedia playback applications [82]. Reproduced by permission of ©2006 IEEE.

	Interruptions	Max queue length (KB)	Transmission rate (KB/s)
DTAIMD	0	1565	95
TCP-1	2	2350	91
TCP-2	12	3440	91
TCP-3	0	3132	88
TCP-4	1	1740	93
TCP-5	0	2329	91
TCP-6	0	1295	87
TCP-7	0	2111	82
TCP-8	7	1860	85
TCP-9	0	1731	84
TCP-10	5	1492	86

the number of interruptions, the maximum queue length (KB) in the receiver end and the average transmission rate (KB/s) of each flow are listed in Table 4.2.

It is observed that the DTAIMD(0.2, 0.875) flow can successfully play back the whole stream without any interruption, while 50% of the TCP flows have 1 to 12 interruptions. In the receiver end, the maximum queue length of the DTAIMD(0.2, 0.875) flow is less than that of 80% of the TCP flows. On the other hand, the average transmission rate of the DTAIMD(0.2, 0.875) flow is close to that of the TCP flows, which indicates that it is indeed TCP-friendly. By repeating the simulation several times, similar results are obtained. It implies that the DTAIMD(0.2, 0.875) algorithm can have better QoS provisioning than TCP for multimedia playback applications.

Without the random *mice* flows, DTAIMD and TCP *elephants* will reach a steady state in which each flow additively increases and multiplicatively decreases its transmission rate periodically. When the random *mice* flows introduce disturbances to the network, TCP flows are more sensitive and have a more prompt response to the disturbances than do DTAIMD(0.2, 0.875) flows, according to the convergence analysis in *Convergent Speed* in Subsection 4.6.1. On the other hand, the number of TCP flows is much larger than that of DTAIMD(0.2, 0.875) flows, which is true in general. Therefore, the DTAIMD flow is immune from the disturbances by the dominant TCP flows and can keep its flow smooth. Although *mice* are more likely to grab bandwidth from TCP flows when they enter the network, TCP flows are more aggressive to grab it back after the *mice* leave the network. In the long term, DTAIMD(0.2, 0.875) can be TCP-friendly, and provide better QoS for multimedia playback applications.

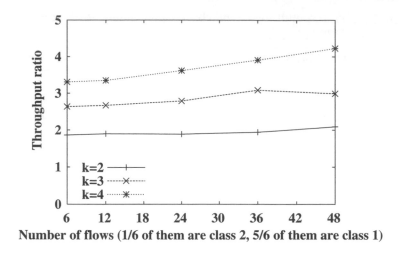

Figure 4.21 Ratio of class 1 and class 2 average throughput [82]. Reproduced by permission of ©2006 IEEE.

Service differentiation

We show the actual performance of utilizing the different AIMD(α, β) pairs to offer service differentiation. Let the multimedia flows (class 2) have a weight k over data flows (class 1). As derived in Subsection 4.5.2, their parameter pairs should satisfy (4.40). Figure 4.21 shows the simulation results of the throughput ratio for multimedia flows over data flows with different k. The x-axis gives the number of flows, and 1/6 of them are multimedia flows. This figure is intended to show that the throughput of class 2 AIMD flows (for multimedia) is k times ($k = 2, 3, 4$) that of the class 1 flows. Since there is no additional upgrade in the core network, it is very effective to offer service differentiation in the Internet by choosing different (α, β) pairs.

In conclusion, according to the TCP-friendly condition derived in Section 4.3, TAIMD flows exhibit satisfactory properties on TCP-friendliness, effectiveness and responsiveness. Furthermore, the DTAIMD-based algorithm can efficiently support multimedia playback applications with service differentiation. These properties meet the design objective and promise a scalable deployment over the Internet.

4.7 Summary

This chapter has introduced the AIMD protocol and derived the necessary and sufficient TCP-friendly condition for AIMD parameters. Analytical and simulation results demonstrate the effectiveness and responsiveness of the AIMD congestion control algorithms. We also have studied the DTAIMD algorithms for multimedia playback applications with differentiated services. Performance studies show that the DTAIMD-based protocol can be truly TCP-friendly over a wide dynamic range and provide better QoS for playback applications. Since AIMD congestion control is simple, compatible and scalable, incremental deployment of the AIMD protocol in the Internet is feasible.

4.8 Problems

1. An important design goal for Internet congestion control is to ensure that competing flows be able to fairly share the bottleneck resources in the long term. A general binomial congestion controller adjusts the *cwnd* W as follows. When no congestion occurs, $W(t + rtt) = W(t) + aW^j(t)$, where $a > 0$; when congestion occurs, $W(t^+) = W(t) - bW^k(t)$, where $0 < b < 1$. Find some parameter sets of (a, b, j, k) (other than $(*, *, 0, 1)$) such that two competing flows under binomial congestion control can converge to fairness, independent of their initial window sizes.

2. As discussed in Subsection 4.2.2, when the received *ack* indicates that three or more out-of-order packets have been received, the AIMD sender will advance W_{left} to $\min(seq_{max} + 1, W_{left} + W)$ (event **E** shown in Figure 4.3). Explain why advancing W_{left} can guarantee that the window is reduced only once per *rtt*.

3. Let two AIMD flows with the same *rtt* share a link with fixed capacity of C packets per *rtt*. The parameter pairs of the two flows are (α_1, β_1) and (α_2, β_2), respectively. Ignore the slow-start phase. At time t_0, their *cwnd*s are W_0 and W_1, and $W_0 + W_1 < C$. What are their *cwnd*s when their window trace enters the overload region for the first time?

4. Let a TCP flow and an AIMD(1, 0.875) flow share a link with fixed capacity 10^3 packets/s and buffer size 100 packets. Without queueing delay, the *rtt*s for both flows are 0.1 s. In steady state, what are the average throughputs of the TCP flow and the AIMD flow, respectively?

5. Use AIMD(α, β) to control the playback application traffic with the consumption rate of C packets per *rtt*. Let the AIMD flow traverse a link of capacity R packets per *rtt*. Denote the amount of data stored during the filling phase B_f, and the amount of data drawn from the receiver buffer during the draining phase B_d, respectively. Derive the ratio of B_f to B_d as given in (4.35).

6. Let one AIMD(0.2, 0.875) flow traverse a single link with a fixed capacity. Without queueing delay, the *rtt*, denoted as τ_0, is constant. All packets are of the same size and the link can transmit R packets per τ_0. The access buffer size is B MSS. Assume that the *rwnd* size of the TCP flow is always larger than its *cwnd*.

(a) What is the maximum achievable window size of the AIMD flow?

(b) What is the average throughput of the flow in steady state (ignore the slow start phase)?

(c) After a sufficiently long time, what is the value of *ssthresh*?

(d) What are the values of the maximum and minimum queueing delay for the AIMD flow?

7. Let one AIMD(0.2, 0.875) flow and a TCP flow share a link. Assume that their *rwnd* sizes are always larger than their *cwnd* sizes, and both flows have an infinite amount of data to be sent. Without queueing delay, the *rtts* of both flows are the same, denoted as $\tau_0 = 50$ ms. The link can transmit 200 packets per second, and the access buffer size is 10 packets. Neglecting the slow start phase, the window sizes of the AIMD and TCP flow at $t = t_0$ are 2 MSS and 4 MSS respectively, and, for each flow, a full window of packets has just been injected into the network back-to-back.

(a) What are the window sizes of the two flows at $t = t_0 + 50$ ms? at $t = t_0 + 100$ ms?

(b) After t_0, when will the first packet loss event occur?

(c) What are the average throughputs of the two flows?

(d) What are the packet loss rates of the two flows?

8. Use NS-2 to investigate the performance of TCP flows and the impact of (α, β) (NS-2 can be downloaded from http://www.isi.edu/nsnam/ns/). The network topology can be set according to the following figure.

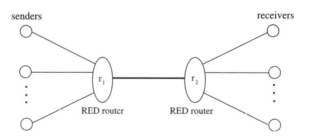

(a) Let $N = 1, 2, 4, 8, 16$ TCP (SACK) flows share the bottleneck link between r_1 and r_2. Packets are of the same size, which is set to 1250 bytes. The bandwidth of each link is 100 Mbps. The propagation delays of link r_1r_2 and of links between r_2 and the receivers are 5 ms. To avoid the *phase effect*, the *rtt* of each flow is made slightly different by setting the propagation delays of the links between the senders and r_1 to be a random variable, uniformly distributed between 4.9 ms and 5.1 ms. The routers are RED-capable, with the following parameter settings in NS-2: thresh_ 100, maxthresh_ 500, linterm_ 10. What are the simulation results of the total throughput, packet loss rate and fairness index?

(b) Change the (α, β) pair of the TCP sender to the following values: (0.2, 0.875), (0.4, 0.765) and (0.6, 0.667) and repeat the above simulations.

9. Let two AIMD(0.5, 0.5) flows, flow I and flow II, share a link with fixed data rate of 10^3 packets/s and buffer size 200 packets. Without queueing delay, the round-trip times for both flows are 0.1 s. The access queue deploys the following management scheme: when the queue length exceeds 100 packets, all packets from flow I will be discarded, and the packets from flow II will be discarded with probability p.

(a) Let $p = 0.5$. In steady state, what are the average throughputs of flow I and flow II, respectively?

(b) If we need to ensure that the average throughput of flow II is twice of that of flow I, explain how you would set the value of p.

10. To download a Web object, Web browser I establishes ten TCP connections, and Web browser II establishes one AIMD(10, 0.5) connection. Compare these two strategies and their impacts on the user perceived quality and network performance.

5

Stability Property and Performance Bounds of the Internet

The first congestion collapse in the Internet was observed in the 1980s, although the Internet was in its infant stage at that time. To solve the problem, Van Jacobson proposed the TCP congestion control algorithm based on the AIMD mechanism in 1988. Since then, the TCP congestion control algorithm has been widely deployed in the end-systems to respond to network congestion indicators and avoid network congestion collapses. On the other hand, the active queue management algorithms, random early detection or random early discard (RED), have been developed and deployed in the intermediate systems to fairly distribute network congestion indicators to all ongoing flows. AIMD and RED are considered important factors in the overwhelming success of the Internet, which has experienced explosive growth over the past two decades, with network speeds almost being doubled every year and new applications emerging quickly. Future Internet will become an even more diversified system. It will contain heterogeneous wireless and wired links with speeds varying from tens of Kbps to tens of Gbps, with flow round-trip delays varying from milliseconds to seconds. It will also support various multimedia applications with different throughput, delay and jitter requirements. An immediate question is whether the AIMD/RED system can be stable, scalable and efficient for the next generation Internet.

Internet stability has been an active research topic since its first congestion collapse was observed. The definitions of stability are listed below, and follow those in [96, 97].

Multimedia Services in Wireless Internet: Modeling and Analysis   Lin Cai, Xuemin Shen and Jon W. Mark
© 2009 John Wiley & Sons, Ltd

Consider dynamic systems with time delay of the following form:

$$\frac{dx}{dt} = f(t, x(t), x(t - \tau_1(t)), \ldots, x(t - \tau_m(t)))$$

where $x \in R^n$, $f: I \times R^n \times R^n \times \cdots \times R^n \to R^n$ is continuous. Let

$$\tau = \sup_{i=1,\ldots,m} \tau_i(t).$$

The trivial solution of the system is said to be:

- *uniformly bounded* if there exists a constant c, for every $a \in (0, c)$, there is $B = B(a) > 0$, such that for any $\xi(t) \in C[[t_0 - \tau, t_0], R^n]$, $\|x(t, t_0, \xi)\| \leq B$ for all $t \geq t_0$ when $\|\xi\| \leq a$;

- *stable* if for every $\epsilon > 0$ and $t_0 \in \mathbb{R}_+$, there exists some $\delta = \delta(t_0, \epsilon) > 0$ such that for any $\xi(t) \in C[[t_0 - \tau, t_0], R^n]$, $\|\xi\| < \delta$ implies $\|x(t, t_0, \xi)\| < \epsilon$ for all $t \geq t_0$;

- *asymptotically stable* if the system is stable and for every $t_0 \in \mathbb{R}_+$, there exists some $\eta = \eta(t_0) > 0$ such that $\lim_{t \to \infty} \|x(t, t_0, \xi)\| = 0$ whenever $\|\xi\| < \eta$;

- *marginally stable* if the system is stable but not asymptotically stable.

With a fluid-flow model of the system, without feedback delay, the AIMD congestion control mechanism, coupled with the RED queue management, can ensure asymptotic stability. However, with a non-negligible feedback delay, the AIMD/RED system may not be asymptotically stable when the delay becomes large and/or when the link capacity becomes large [98].

On the other hand, the Internet is a very dynamic system, and can tolerate some transient congestion events. In fact, TCP controlled flows aggressively probe for available bandwidth in the network, and create transient congestions. Even if the system is not asymptotically stable, so long as the end-systems do not overshoot the available bandwidth too severely, the overall system efficiency can still be very high, and the packet loss rate and queueing delay can still be well bounded. Therefore, it is critical to investigate whether the network can operate at states away from the desired equilibrium state, and what the theoretical bounds of the system are.

In this chapter, we first introduce the fluid-flow model for the AIMD/RED system. Based on the model, the system without feedback delay is proved to be asymptotically stable [99]. For a given number of AIMD flows sharing the link, the theoretical bounds of the system, including the congestion window size of the flows and the queue length of the intermediate systems, can be obtained. With clearly defined bounds, the system is considered marginally stable. The derived theoretical bounds shed light on which system parameters contribute to high oscillations of the system and how to choose system parameters to ensure system efficiency with bounded delay and loss. The theorems given can also help to predict the system performance for the future Internet with higher data rate links and supporting heterogeneous flows.

The remainder of this chapter is organized as follows. Section 5.1 introduces the model of the generalized AIMD/RED system. Section 5.2 presents the asymptotic stability property of the generalized AIMD/RED system with delay-free marking [99]. Section 5.3 studies the boundedness of AIMD/RED systems with feedback delay. Numerical results with Matlab and simulation results using NS-2 are presented to validate the derived bounds, and the impacts of different system parameters on the system performance are discussed. Concluding remarks are presented in Section 5.4.

5.1 A Fluid-flow Model of the AIMD/RED System

A stochastic model of TCP behaviors is developed using fluid-flow[1] and stochastic differential equation analysis in [101]. Simulation results have demonstrated that this model accurately captures the dynamics of TCP. The fluid-flow model can be extended for general AIMD(α, β) congestion control discussed in the previous chapter: the window size is increased by α packets per *rtt* if no packet loss occurs; otherwise, it is reduced to β times its current value.

Assuming all AIMD-controlled flows have the same (α, β) parameter pair and round-trip delay, the AIMD fluid model is related to the *ensemble averages* of key network variables [101, 102], and can be described by the following coupled, nonlinear differential equations:

$$\frac{dW(t)}{dt} = \frac{\alpha}{R(t)} - \frac{2(1-\beta)}{1+\beta} W(t) \frac{W(t-R(t))}{R(t-R(t))} p(t-R(t)),$$

$$\frac{dq(t)}{dt} = \begin{cases} \dfrac{N(t) \cdot W(t)}{R(t)} - C, & q > 0 \\ \left\{ \dfrac{N(t) \cdot W(t)}{R(t)} - C \right\}^+, & q = 0, \end{cases} \tag{5.1}$$

where $\{a\}^+ = \max\{a, 0\}$, $\alpha > 0$, $\beta \in [0, 1]$; $W \geq 1$ is the AIMD window size (packets per *rtt*), and $q \in [0, q_{max}]$ is the queue length (packets) at time t. W and q in the fluid-flow model can approximate the ensemble averages of the flow's congestion window size and queue length, respectively, in a real system. $R(t) = q(t)/C + T_p(s)$ is the round-trip time, where C is the link capacity (packets per second) and T_p is the deterministic round-trip delay. At time t, $N(t)$ denotes the number of AIMD flows, and $p(t)$ denotes the probability of a packet being dropped or marked by an intermediate system.

---

[1]The concept of fluid in data networks is based on the assumption that in high capacity networks, most important dynamics do not depend on how individual packets are processed, but rather on how aggregates of packets are processed. A good criterion to test the applicability of the assumption is that packet size should be a small fraction of typical buffer capacities in the network [100].

The first differential equation of system (5.1) describes the AIMD(α, β) window control dynamic. $\alpha/R(t)$ represents the window's additive increase, whereas $2(1 - \beta)W(t)/(1 + \beta)$ represents the window's multiplicative decrease in response to packet dropping or marking probability p. Since the AIMD flow's window size in a practical system oscillates between βW_{max} and W_{max}, its average window size W over a cycle[2] is $(1 + \beta)W_{max}/2$. Each time, the window size is decreased by $(1 - \beta)W_{max} = 2(1 - \beta)W/(1 + \beta)$. The second equation models the bottleneck queue length as simply an accumulated difference between packet arrival rate NW/R and link capacity C. $\{\cdot\}^+$ in the model guarantees that the queue length is non-negative.

It should be noted that, in the fluid-flow model, q and W are positive and bounded quantities which approximate the *ensemble* averages of the queue length and the window size in a practical system, respectively. In ergodic systems, ensemble average equals time average. The values of q and W in the fluid-flow model can be used to predict its time average over a cycle in a practical system. Given the AIMD window size oscillating between βW_{max} and W_{max} in a cycle, the average duration of a cycle equals $2(1 - \beta)WR/[(1 + \beta)\alpha]$.

Consider the deployment of the RED scheme in the system (5.1). The packet dropping or marking probability, p, is determined by the average queue length q_{act}:

$$p = \begin{cases} 0, & 0 \leq q_{act} \leq \min_{th} \\ K_p(q_{act} - \min_{th}), & \min_{th} < q_{act} \leq \max_{th} \\ 1, & q_{act} > \max_{th}, \end{cases} \qquad (5.2)$$

where $K_p > 0$ is an important RED queue parameter. When $q_{act} \leq \min_{th}$, $dW(t)/dt = \alpha/R$, the window size of AIMD flows will keep increasing and will not converge to any value. Thus, in the following, we will discuss the stability of this system when $q_{act} > \min_{th}$. Without loss of generality, let $q(t) = q_{act}(t) - \min_{th}$. In addition, since the queue behaves in the same way as a Drop-Tail queue once q_{act} exceeds $\max_{th}$, we choose $\max_{th}$ to be sufficiently large such that $K_p(\max_{th} - \min_{th}) = 1$.

The system in (5.1) is a generalized AIMD/RED congestion control model: if we choose $\alpha = 1$, $\beta = 0.5$, (5.1) is equivalent to the TCP/RED model in [101].

5.2   Stability and Fairness Analysis with Delay-free Marking

5.2.1   Stability of the homogeneous AIMD/RED system

With the fluid-flow model (5.1), we assume that the traffic load (N AIMD flows) is time-invariant, i.e., $N(t) = N$, and the round-trip time of each flow is constant,

---

[2]A cycle in this chapter is defined as the interval between two time instants at which the flow reduces its congestion window size consecutively.

$R(t) = R$. In the case of delay-free marking, i.e., $p = K_p q(t)$, the original model (5.1) can be written as a closed-loop dynamical system:

$$\frac{dW(t)}{dt} = \frac{\alpha}{R} - \frac{2(1-\beta)}{1+\beta} W(t) \frac{W(t)}{R} K_p q(t),$$

$$\frac{dq(t)}{dt} = \begin{cases} \dfrac{N \cdot W(t)}{R} - C, & q > 0 \\[2mm] \left\{ \dfrac{N \cdot W(t)}{R} - C \right\}^{+}, & q = 0. \end{cases} \tag{5.3}$$

For a single-bottleneck system, the equilibrium point (W_0^*, q_0^*) for (5.3) is given by

$$W_0^* = \frac{RC}{N}; \qquad q_0^* = \frac{\alpha(1+\beta)N^2}{2(1-\beta)R^2C^2K_p}. \tag{5.4}$$

Remarks. At equilibrium, the total arrival rate equals the link capacity, so the link bandwidth can be fully utilized. If the window size is larger than W_0^*, the queue will build up, which results in a longer queueing delay; if the window size is less than W^*, the traffic load is smaller than the link capacity, so the network resources are not fully utilized. In conclusion, the equilibrium point is also the most desired operating point of the system.

With the transformed variables $\tilde{W} := W - W_0^*$, $\tilde{q} := q - q_0^*$, (5.3) becomes

$$\dot{\tilde{W}}(t) = -\frac{2(1-\beta)}{1+\beta} \frac{(\tilde{W}(t)+W_0^*)^2}{R} K_p \tilde{q}(t) - \frac{2(1-\beta)}{1+\beta} \frac{\tilde{W}^2(t)+2\tilde{W}(t)W_0^*}{R} K_p q_0^*,$$

$$\dot{\tilde{q}}(t) = \frac{N}{R} \cdot \tilde{W}(t). \tag{5.5}$$

The equilibrium point of (5.5) is $(\tilde{W}^*, \tilde{q}^*) = (0, 0)$.

We construct the positive-definite Lyapunov function,

$$V(\tilde{W}, \tilde{q}) = \frac{(1+\beta)N^3}{2(1-\beta)R^2C^2} \cdot \tilde{W}^2(t) + \frac{1}{2} K_p \tilde{q}^2(t),$$

which is used to derive the following theorem.

Theorem 2 *The equilibrium point of (5.3) is asymptotically stable for all $K_p > 0$.*

The proof of Theorem 2 is omitted, and we will prove a more general theorem (Theorem 3) in the next subsection.

The block diagram of the AIMD/RED system, from a control theory point of view, is shown in Figure 5.1 [103]. By a suitable control law, we can relate the output q with the input p, which makes the original open loop systems into a closed loop control system to achieve asymptotic stability.

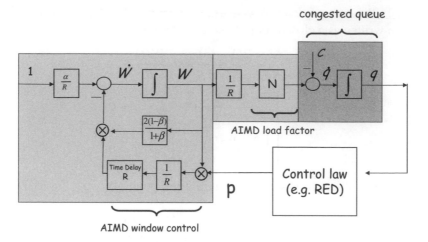

Figure 5.1 Block diagram of generalized AIMD/RED system [99]. Reprinted from Wang, L., Cai, L., Liu, X. and Shen, X. (2007) Stability and TCP-friendliness of AIMD/RED systems with feedback delays. *Elsevier Journal of Computer Networks*, 51(15):4475–4491.

5.2.2 Stability of the heterogeneous AIMD/RED system

In the previous subsection, we have discussed the stability property of the homogeneous-flow system when there is only one type of flow with the parameter pair (α, β). To support heterogeneous multimedia applications, we study the system with heterogeneous flows, i.e., there are two or more types of flows with the parameter pairs (α_1, β_1), (α_2, β_2), ..., (α_m, β_m).

First, we consider the case when there are two different heterogeneous flows: W_I, whose *rtt* is R_1, and W_{II}, whose *rtt* is R_2, with the parameters (α_1, β_1), (α_2, β_2), respectively. The number of W_I flows is N_1, and that of W_{II} flows is N_2. Then the corresponding mathematical model has the following form:

$$\frac{dW_I(t)}{dt} = \frac{\alpha_1}{R_1} - \frac{2(1-\beta_1)}{1+\beta_1} \cdot \frac{W_I^2(t)}{R_1} \cdot K_p q(t),$$

$$\frac{dW_{II}(t)}{dt} = \frac{\alpha_2}{R_2} - \frac{2(1-\beta_2)}{1+\beta_2} \cdot \frac{W_{II}^2(t)^2}{R_2} \cdot K_p q(t),$$

$$\frac{dq(t)}{dt} = \begin{cases} \dfrac{N_1 W_I(t)}{R_1} + \dfrac{N_2 W_{II}(t)}{R_2} - C, & q > 0 \\[2mm] \left\{ \dfrac{N_1 W_I(t)}{R_1} + \dfrac{N_2 W_{II}(t)}{R_2} - C \right\}^+, & q = 0. \end{cases} \tag{5.6}$$

The equilibrium points $(W_I^*, \; W_{II}^*, \; q_0^*)$ of (5.6) can be obtained as

$$W_I^* = \frac{R_1 R_2 C}{R_2 N_1 + (\alpha_2(1-\beta_1)(1+\beta_2)/\alpha_1(1+\beta_1)(1-\beta_2))^{1/2} \cdot R_1 N_2},$$

$$W_{II}^* = \frac{R_1 R_2 C}{(\alpha_1(1+\beta_1)(1-\beta_2)/\alpha_2(1-\beta_1)(1+\beta_2))^{1/2} \cdot R_2 N_1 + R_1 N_2},$$

$$q_0^* = \frac{\alpha_1(1+\beta_1)[R_2 N_1 + (\alpha_2(1-\beta_1)(1+\beta_2)/\alpha_1(1+\beta_1)(1-\beta_2))^{1/2} R_1 N_2]^2}{2 R_1^2 R_2^2 C^2 K_p (1-\beta_1)}^{1/2}.$$

$$(5.7)$$

With the transformed variables $\tilde{W}_I(t) := W_I(t) - W_I^*, \quad \tilde{W}_{II}(t) := W_{II}(t) - W_{II}^*$ and $\tilde{q}(t) := q(t) - q_0^*$, (5.6) becomes

$$\dot{\tilde{W}}_I(t) = -\frac{2(1-\beta_1)}{1+\beta_1} \frac{(\tilde{W}_I(t) + W_I^*)^2}{R_1} K_p \tilde{q}(t)$$

$$- \frac{2(1-\beta_1)}{1+\beta_1} \frac{\tilde{W}_I^2(t) + 2 W_I^* \tilde{W}_I(t)}{R_1} K_p q_0^*,$$

$$\dot{\tilde{W}}_{II}(t) = -\frac{2(1-\beta_2)}{1+\beta_2} \frac{(\tilde{W}_{II}(t) + W_{II}^*)^2}{R_2} K_p \tilde{q}(t)$$

$$- \frac{2(1-\beta_2)}{1+\beta_2} \frac{\tilde{W}_{II}^2(t) + 2 W_{II}^* \tilde{W}_{II}(t)}{R_2} K_p q_0^*,$$

$$\dot{\tilde{q}}(t) = \frac{N_1 \cdot \tilde{W}_I(t)}{R_1} + \frac{N_2 \cdot \tilde{W}_{II}(t)}{R_2}. \qquad (5.8)$$

The equilibrium point of (5.8) is then $(\tilde{W}_I^*, \; \tilde{W}_{II}^*, \; \tilde{q}_0^*) = (0, 0, 0)$.

With (5.8), choose the following positive-definite Lyapunov function:

$$V(\tilde{W}_I(t), \; \tilde{W}_{II}(t), \; \tilde{q}(t))$$

$$= \frac{(1+\beta_1)N_1}{2(1-\beta_1)W_I^{*2}} \cdot \tilde{W}_I^2(t) + \frac{(1+\beta_2)N_2}{2(1-\beta_2)W_{II}^{*2}} \cdot \tilde{W}_{II}^2(t) + K_p \tilde{q}^2(t).$$

Then,

$$\dot{V} = \frac{(1+\beta_1)N_1}{(1-\beta_1)W_I^{*2}} \tilde{W}_I(t) \dot{\tilde{W}}_I(t) + \frac{(1+\beta_2)N_2}{(1-\beta_2)W_{II}^{*2}} \tilde{W}_{II}(t) \dot{\tilde{W}}_{II}(t) + 2 K_p \tilde{q}(t) \dot{\tilde{q}}(t)$$

$$= -\frac{2 N_1 K_p}{W_I^{*2} R_1} \tilde{W}_I^2(t)(\tilde{W}_I(t) + 2 W_I^*)(\tilde{q}(t) + q_0^*)$$

$$- \frac{2 N_2 K_p}{W_{II}^{*2} R_2} \tilde{W}_{II}^2(t)(\tilde{W}_{II}(t) + 2 W_{II}^*)(\tilde{q}(t) + q_0^*)$$

$$\leq 0.$$

From the physics constraint point of view, the positive-definite Lyapunov function is the total energy function of the system, i.e., the sum of kinetic and potential energy. Here, $\dot{V} \leq 0$, since $\tilde{W}_I(t) + 2W_I^* > 0$, $\tilde{W}_{II}(t) + 2W_{II}^* > 0$ and $\tilde{q}(t) + q_0^* \geq 0$, which means the energy of the system is nonincreasing. Thus, we prove that the equilibrium point is stable. To conclude asymptotic stability, we first consider the set of states where $\dot{V} = 0$:

$$\mathcal{M} := \{(\tilde{W}_I, \tilde{W}_{II}, \tilde{q}) : \dot{V} = 0\}$$
$$= \{(\tilde{W}_I, \tilde{W}_{II}, \tilde{q}) : \tilde{W}_I = \tilde{W}_{II} = 0 \text{ or } \tilde{q} = -q_0^*\}.$$

By LaSalle's Invariance Principle [97], trajectories of (5.8) converge to the largest invariant set contained in $\mathcal{M}$. We will then prove that the only invariant set contained in $\mathcal{M}$ is the equilibrium point $(0, 0, 0)$. If $(\tilde{W}_I, \tilde{W}_{II}, \tilde{q})$ is equal to $(0, 0, \tilde{q})$ or $(\tilde{W}_I, \tilde{W}_{II}, -q_0^*)$, by using (5.8), we can conclude that $(\tilde{W}_I(t^+), \tilde{W}_{II}(t^+), \tilde{q}(t^+))$ is not in $\mathcal{M}$, which implies that no trajectory can stay in $\mathcal{M}$, other than the point $(0, 0, 0)$. Therefore, asymptotic stability is obtained, which we summarize as follows:

Theorem 3 *For any $K_p > 0$, the equilibrium point of (5.8) is asymptotically stable for any positive pairs (α_1, β_1), (α_2, β_2) and any positive R_1, R_2.*

We can also extend our results to the case when more than two heterogeneous flows exist in the same system. Suppose that there are M types of AIMD flows with parameters (α_1, β_1), (α_2, β_2), ..., (α_m, β_m) sharing the resources, with the number $N_1, N_2, \ldots, N_m$ and different $rtts$ $R_1, R_2, \ldots, R_m$ respectively, then those flows can be mathematically modeled as

$$\frac{dW_I(t)}{dt} = \frac{\alpha_1}{R_1} - \frac{2(1-\beta_1)}{1+\beta_1} \cdot \frac{W_I(t)^2}{R_1} \cdot K_p q(t),$$

$$\frac{dW_{II}(t)}{dt} = \frac{\alpha_2}{R_2} - \frac{2(1-\beta_2)}{1+\beta_2} \cdot \frac{W_{II}(t)^2}{R_2} \cdot K_p q(t),$$

$$\cdots\cdots\cdots\cdots$$

$$\frac{dW_M(t)}{dt} = \frac{\alpha_m}{R_m} - \frac{2(1-\beta_m)}{1+\beta_m} \cdot \frac{W_M(t)^2}{R_m} \cdot K_p q(t),$$

$$\frac{dq(t)}{dt} = \begin{cases} \sum_{i=1}^{m} \frac{N_i W_i(t)}{R_i} - C, & q > 0 \\ \left\{\sum_{i=1}^{m} \frac{N_i W_i(t)}{R_i} - C\right\}^+, & q = 0. \end{cases} \tag{5.9}$$

With (5.9), we choose a positive-definite Lyapunov function as

$$V(\tilde{W}_I(t), \tilde{W}_{II}(t), \ldots, \tilde{W}_M(t), \tilde{q}(t))$$

$$= \frac{(1+\beta_1)N_1}{2(1-\beta_1)W_I^{*2}} \cdot \tilde{W}_I^2(t) + \frac{(1+\beta_2)N_1}{2(1-\beta_2)W_{II}^{*2}} \cdot \tilde{W}_{II}^2(t)$$

$$+ \cdots + \frac{(1+\beta_m)N_m}{2(1-\beta_m)W_M^{*2}} \cdot \tilde{W}_M^2(t) + K_p \tilde{q}^2(t),$$

where $\tilde{W}_i(t)$, $i = 1, 2, \ldots, M$ and $\tilde{q}(t)$ have the same meaning as in (5.8). Then,

$$\dot{V} = \frac{(1+\beta_1)N_1}{(1-\beta_1)W_I^{*2}} \tilde{W}_I \dot{\tilde{W}}_I + \frac{(1+\beta_2)N_2}{(1-\beta_2)W_{II}^{*2}} \tilde{W}_{II} \dot{\tilde{W}}_{II}$$

$$+ \cdots + \frac{(1+\beta_m)N_M}{(1-\beta_m)W_M^{*2}} \tilde{W}_M \dot{\tilde{W}}_M + 2K_p \tilde{q} \dot{\tilde{q}}$$

$$= -\frac{2N_1 K_p}{W_I^{*2} R_1} \tilde{W}_I^2 (\tilde{W}_I + 2W_I^*)(\tilde{q} + q_0^*)$$

$$- \cdots - \frac{2N_m K_p}{W_M^{*2} R_m} \tilde{W}_M^2 (\tilde{W}_M + 2W_M^*)(\tilde{q} + q_0^*)$$

$$\leq 0.$$

We can obtain its asymptotic stability by applying LaSalle's Invariance Principle, and thus have the following theorem:

Theorem 4 *For any $K_p > 0$, the equilibrium point of system (5.9) is asymptotically stable for any positive pairs (α_1, β_1), (α_2, β_2), $\ldots$, (α_m, β_m) and any positive $R_1, R_2, \ldots, R_m$.*

5.2.3 TCP-friendliness and differentiated services

For two competing AIMD flows, from (5.7), we can also get the relationship between W_I^* and W_{II}^* as follows:

$$\frac{W_I^*}{W_{II}^*} = \left[\frac{\alpha_1(1+\beta_1)(1-\beta_2)}{\alpha_2(1-\beta_1)(1+\beta_2)} \right]^{1/2}. \tag{5.10}$$

This means that the ratio of W_I^* and W_{II}^* depends only on the choices of (α_1, β_1) and (α_2, β_2), regardless of the traffic loads in the network and their initial states. Consequently, for AIMD(α, β) flows to be TCP-friendly, i.e., co-existing TCP and AIMD flows obtain the same share of bottleneck bandwidth, the necessary and sufficient condition is

$$\alpha = \frac{3(1-\beta)}{1+\beta}. \tag{5.11}$$

Equation (5.10) indicates that we can easily adjust the AIMD parameters of the end-systems to provide differentiated services according to different QoS requirements. For instance, let the throughput of an $\text{AIMD}(\alpha_2, \beta_2)$ flow be k times that of an $\text{AIMD}(\alpha_1, \beta_1)$ flow, the AIMD parameter pairs should satisfy

$$\frac{\alpha_2}{\alpha_1} = \frac{k^2(1+\beta_1)(1-\beta_2)}{(1-\beta_1)(1+\beta_2)}. \qquad (5.12)$$

Remark. The TCP-friendly condition (5.11) and the condition to provide differentiated services (5.12) derived here are the same as (4.12) and (4.40) derived in the previous chapter, although different approaches are used in these two chapters.

5.2.4 Numerical results

The traces of the average window size and queue length of 100 TCP ($\alpha = 1$, $\beta = 0.5$) flows and 100 AIMD(0.2, 0.875) flows are given in Figures 5.2 and 5.3, respectively. The parameters used are $C = 100\,000$ packets/s, $R = 100$ ms, $K_p = 0.0001$ and $\min_{th} = 200$ packets. For the TCP-friendliness, let 100 TCP flows and 24 AIMD(0.2, 0.875) flows share the bottleneck; the numerical results obtained using Matlab are shown in Figure 5.4. It can be seen that when the flows in the network possess the same (α, β) parameter pair, the ensemble averages of the window size and the bottleneck queue length converge to certain values, i.e., the equilibrium points we derived in the previous analysis. When TCP and AIMD(0.2, 0.875) flows co-exist, they will fairly share the link capacity in steady state, since (0.2, 0.875) satisfies the TCP-friendly condition (5.11). Thus, the numerical results validate the theorems.

Furthermore, from Figures 5.2 and 5.3, with a smaller value of α and a larger value of β, it takes a longer time for the system to converge to steady state, and the link utilization during the transient stage is low; however, in steady state, the oscillation amplitudes of the instantaneous window size and queue length are smaller. In other words, with a smaller value of α and a larger value of β, the queueing delay jitter is smaller and the link utilization in steady state is higher, which are desired for supporting time-sensitive multimedia applications.

5.3 Boundedness of the Homogeneous-flow AIMD/RED System with Time Delay

When feedback delay is considered, AIMD/RED systems may not always be asymptotically stable. Using the fluid-flow model, sufficient conditions for the asymptotic stability of AIMD/RED systems with feedback delays have been derived in the previous section [99]. It has been demonstrated in [98] that an AIMD/RED system becomes (asymptotically) unstable with the increase of round-trip delays of the system. In this section, we show that even though the system may become (asymptotically) unstable because of the effects of time delay, its window size and

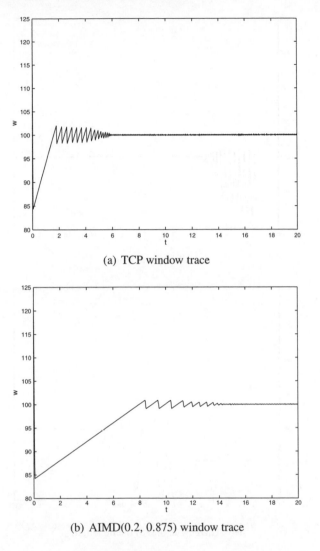

(a) TCP window trace

(b) AIMD(0.2, 0.875) window trace

Figure 5.2 Window trace [99]. Reprinted from Wang, L., Cai, L., Liu, X. and Shen, X. (2007) Stability and TCP-friendliness of AIMD/RED systems with feedback delays. *Elsevier Journal of Computer Networks*, 51(15):4475–4491.

queue length are still bounded, and in most cases the upper bounds are close to their equilibria.

We study the delayed homogeneous AIMD system defined by (5.1) with RED defined by (5.2). We set $\min_{th} = 0$ in RED for presentation simplicity, and assume that the traffic load (i.e., the number of AIMD flows) is time-invariant, i.e., $N(t) = N$. With ever-increasing link capacity and appropriate congestion control mechanism,

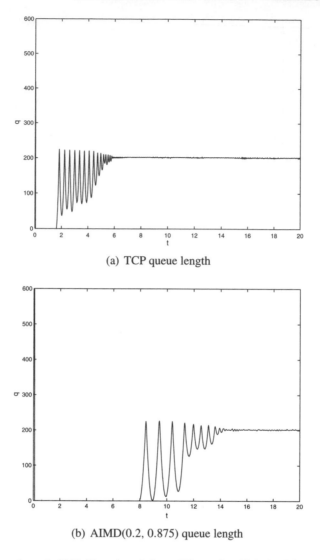

(a) TCP queue length

(b) AIMD(0.2, 0.875) queue length

Figure 5.3 Queue length [99]. Reprinted from Wang, L., Cai, L., Liu, X. and Shen, X. (2007) Stability and TCP-friendliness of AIMD/RED systems with feedback delays. *Elsevier Journal of Computer Networks*, 51(15):4475–4491.

variation of queueing delays becomes relatively small compared with propagation delays. In [104], it is shown that variations in *rtt* due to queueing delay variation help to stabilize the TCP/RED system. In light of this, we derive upper and lower bounds of AIMD/RED systems assuming *rtt* to be constant. These results will be a good approximation if *rtt* is slightly time-varying. We thus ignore the effect of delay jitter

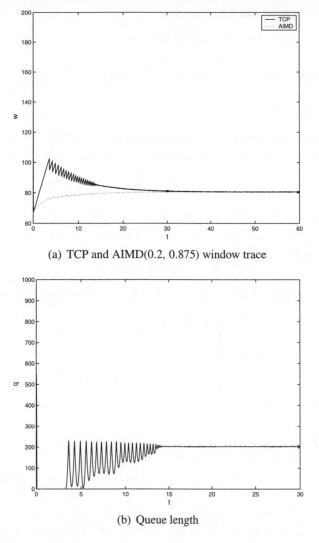

(a) TCP and AIMD(0.2, 0.875) window trace

(b) Queue length

Figure 5.4   TCP-friendliness [99]. Reprinted from Wang, L., Cai, L., Liu, X. and Shen, X. (2007) Stability and TCP-friendliness of AIMD/RED systems with feedback delays. *Elsevier Journal of Computer Networks*, 51(15):4475–4491.

on the round-trip time and assume that the round-trip time of each flow is constant, $R(t) = R$.

Notice that the AIMD/RED system defined by (5.1) and (5.2) is described by delayed differential equations, with initial conditions given by $1 \leq W(t) \leq W^*$ and $0 \leq q(t) \leq q^*$ on the interval $t \in [-R, \ 0]$. According to (5.1), it is also reasonable that we let $\dot{W}(t) \leq \alpha/R$ for $t \in [-R, \ 0]$.

5.3.1   Upper bound on window size

Theorem 5 *Let UB > 0 be the largest real root of*

$$UB \cdot (UB - \alpha) \cdot \left(UB - \frac{R \cdot C}{N} - \alpha \right)^2 = \frac{\alpha^2 (1 + \beta)}{(1 - \beta) N K_p}.$$

Then $W(t) \leq UB$ for $t \geq 0$.

   Proof. With (5.1) and (5.2), we note that $\dot{W} \leq \alpha / R$ for $t \geq 0$, since $W(t) \geq 1$ and $q(t) \geq 0$. For $\tau > 0$, taking integration on both sides from $t - \tau$ to t gives

$$W(t) - W(t - \tau) \leq \frac{\alpha}{R} \cdot \tau \quad \text{for } t \geq 0. \tag{5.13}$$

   UB (> 0) in the theorem is an upper bound of $W(t)$ for $t \geq 0$, i.e., if $W(t) = UB$ for some $t = t_1 \geq 0$, then $\dot{W}(t_1) \leq 0$. With (5.13) and $W(t_1) = UB$, and letting $\tau = R$ and $t = t_1$, we have

$$W(t_1 - R) \geq UB - \alpha. \tag{5.14}$$

Note that when $\tau \in [R, \ aR]$, $W(t_1 - \tau) \geq UB - a \cdot \alpha$ for any real number $a > 1$.
   Consider

$$\dot{q}(t) = \begin{cases} \dfrac{N \cdot W(t)}{R} - C, & q > 0 \\[2ex] \left\{ \dfrac{N \cdot W(t)}{R} - C \right\}^+, & q = 0. \end{cases}$$

Taking integration on both sides from $t_1 - aR$ to $t_1 - R$, we have

$$\int_{t_1 - aR}^{t_1 - R} \dot{q}(s) ds \geq \frac{N}{R} \int_{t_1 - aR}^{t_1 - R} W(s) ds - (a - 1) R \cdot C$$

$$\geq N \cdot (a - 1) \cdot (UB - a \cdot \alpha) - (a - 1) RC,$$

which implies

$$q(t_1 - R) \geq [N \cdot (UB - a \cdot \alpha) - R \cdot C] \cdot (a - 1), \tag{5.15}$$

since $q(t) \geq 0$.
   Taking $f(a) = (a - 1) \cdot [N \cdot (UB - a \cdot \alpha) - R \cdot C]$ and computing the maximum value of $f(a)$ by letting $f'(a) = 0$, we have $a = (N \cdot UB + R \cdot C + N \cdot \alpha) / (2 \alpha N)$ and

$$f(a) = N(UB - R \cdot C / N - \alpha)^2 / (2\alpha). \tag{5.16}$$

Therefore, it follows from (5.14), (5.15) and (5.16) that, $\dot{W}(t_1) \leq 0$ since UB satisfies

$$\frac{N \cdot UB \cdot (UB - \alpha) \cdot (UB - R \cdot C / N - \alpha)^2}{2\alpha} = \frac{\alpha(1 + \beta)}{2(1 - \beta) K_p}, \tag{5.17}$$

which implies $W(t) \leq UB$ for $t \geq 0$.   □

If all AIMD flows are TCP-friendly, the (α, β) pair satisfies the TCP-friendly condition $\alpha = 3(1 - \beta)/(1 + \beta)$. Thus (5.17) can be manipulated to yield

$$UB \cdot (UB - \alpha) \cdot (UB - R \cdot C/N - \alpha)^2 = \frac{3\alpha}{NK_p}. \tag{5.18}$$

By the continuity property of $UB \cdot (UB - \alpha) \cdot (UB - R \cdot C/N - \alpha)^2$ and the fact that the RHS of (5.17) is always greater than zero, we can conclude that the largest root of (5.17) must be greater than $R \cdot C/N + \alpha$, where $R \cdot C/N$ is the equilibrium value of the window size for the AIMD/RED system. Therefore, the oscillation of the window size about its equilibrium value will increase with the increment of α and the decrement of K_p. In addition, the upper bound UB itself will increase with the increment of $R \cdot C$ and α, and the decrement of N and K_p.

It is also noted that the upper bound derived in Theorem 5 is a global one for the time t, i.e., the window size $W(t)$ will not go above UB for any $t > t_1$. If we assume, instead, that there exists $t_1' > t_1$ and $\Delta W > 0$, such that $W(t_1') = UB + \Delta W$, then there must be some $\tau' \in (0, \; t_1' - t_1)$ such that $W(t_1' - \tau') = UB$ and $\dot{W}(t_1' - \tau') > 0$. However, similar to the proof of Theorem 5, we have $\dot{W}(t_1' - \tau') \leq 0$, which is a contradiction. Therefore, the window size is upper bounded by UB for any $t \geq 0$.

5.3.2 Lower bound on window size and upper bound on queue length

In the previous subsection, we proved that the AIMD window size $W(t)$ is bounded from above, and an upper bound, UB, is defined by (5.17). In this subsection, we show that the window size is also bounded from below while the queue length is upper bounded.

Theorem 6 *Define*

$$A := \frac{\alpha}{R} - \frac{2(1 - \beta)}{1 + \beta} \frac{UB^2}{R}$$

and let $LB_1 > 0$ be the root of

$$LB_1 \cdot (LB_1 - AR) = \frac{\alpha(1 + \beta)}{2(1 - \beta)},$$

then $W(t) \geq LB_1$ for $t \geq 0$.

Proof. From Theorem 5, $W(t) \leq UB$ for $t \geq 0$, which implies

$$\dot{W}(t) \geq \frac{\alpha}{R} - \frac{2(1 - \beta)}{1 + \beta} \frac{UB^2}{R} =: A.$$

It can be seen from the definition of UB that $A < 0$. We show that $LB_1 > 0$ is the lower bound of $W(t)$ for $t \geq 0$, i.e., if $W(t) = LB_1$ at time $t = t_2 \geq 0$, then $\dot{W}(t_2) \geq 0$.

Taking integration on both sides from $t_2 - R$ to t_2 gives $W(t_2 - R) \leq W(t_2) - AR = LB_1 - AR$.

Since the dropping/marking probability $p(t) = K_p \cdot q(t) \le 1$ for all t, then

$$\dot{W}(t_2) \ge \frac{\alpha}{R} - \frac{2(1-\beta)}{1+\beta} \frac{LB_1 \cdot (LB_1 - AR)}{R}.$$

Therefore, $\dot{W}(t_2) \ge 0$ since LB_1 satisfies

$$LB_1 \cdot (LB_1 - AR) = \frac{\alpha(1+\beta)}{2(1-\beta)}, \qquad (5.19)$$

which implies $W(t) \ge LB_1$ for $t \ge 0$.   □

Notice that LB_1 in Theorem 6 is the lower bound of $W(t)$ for all $t \ge 0$, which is a global one. By similar analysis to the upper bound of the window size UB, it is easy to check that the window size $W(t)$ will not go below LB_1 for any $t > t_2$. However, the value of LB_1 is actually very small since $\alpha(1+\beta)/(2(1-\beta))$ is fairly small compared with $-AR$. Therefore, the global lower bound does not provide much information about the performance of AIMD/RED systems.

Since the window size oscillates around its equilibrium in steady state, the amplitude of the oscillation is more important than the global lower bound. Next, we will show the local lower bound of the window size after the first time it reaches the peak value at time instant t_1. This local lower bound is more useful for understanding the performance of AIMD/RED systems.

Theorem 7 *Define T_1 and UQ as*

$$T_1 = \frac{UB - \frac{R \cdot C}{N}}{\frac{2(1-\beta)}{1+\beta} \cdot \frac{C \cdot K_p}{N} \cdot \left[\frac{R \cdot C}{N} \Delta q + \Delta W(q_0^* + \Delta q) \right]},$$

$$UQ = \inf_{\substack{\Delta q > 0, \\ \Delta W \in [0,\ UB - R \cdot C/N]}} \left\{ (q_0^* + \Delta q) + \left(\frac{N}{R} \cdot UB - C \right) \cdot (T_1 + R) \right\},$$

where UB is as defined in Theorem 5. Let $LB_2 > 0$ satisfy

$$LB_2 \cdot \left(LB_2 + \frac{2(1-\beta)}{1+\beta} \cdot UB^2 \cdot K_p \cdot UQ - \alpha \right) \cdot K_p \cdot UQ = \frac{\alpha(1+\beta)}{2(1-\beta)},$$

then $q(t) \le UQ$ for $t \ge 0$ and $W(t) \ge LB_2$ for $t \ge t_1$.

Proof. We first derive the upper bound of $q(t)$ for $t \ge 0$. At time instant $t = t_1$, $W(t)$ reaches its peak value. To get a loose upper bound of $q(t)$, we assume that $W(t)$ does not decrease for some time after t_1, and thus $q(t)$ increases at the rate $\frac{N}{R} UB - C$. If t_1' is chosen such that $q(t_1') = q^* + \Delta q$ with $\Delta q > 0$, then $W(t)$ decreases from t_1' while $q(t)$ keeps increasing until t_2 such that $\dot{q}(t_2) = 0$ (i.e., $W(t_2) = R \cdot C/N$). Therefore, $q(t_2)$ is the local maximum value of $q(t)$. It is noted that this estimate of $q(t)$ might be greater than the true maximum value of $q(t)$ since $W(t)$ may not stay at its peak value after t_1, and $q(t)$ may increase after t_1 at a rate less than $\frac{N}{R} UB - C$.

From the above analysis, for $t \in [t_1', t2]$, $\dot{q}(t) \leq (N/R) \cdot UB - C$. Thus,

$$\int_{t_1'}^{t_2} \dot{q}(s)ds \leq \left(\frac{N}{R} \cdot UB - C\right) \cdot (t_2 - t_1'),$$

which implies

$$q(t_2) \leq q(t_1') + \left(\frac{N}{R} \cdot UB - C\right) \cdot (t_2 - t_1')$$

$$= (q_0^* + \Delta q) + \left(\frac{N}{R} \cdot UB - C\right) \cdot (t_2 - t_1'). \tag{5.20}$$

To estimate the length of the interval $[t_1', t_2]$, for $t \in [t_1' + R, t_2]$, it follows from the analysis above that

$$W(t) \geq W(t_2) = \frac{R \cdot C}{N},$$

$$q(t - R) \geq q(t_1') = q_0^* + \Delta q,$$

$$W(t - R) \geq W(t_2 - R) = \frac{R \cdot C}{N} + \Delta W,$$

for some $\Delta q > 0$ and $\Delta W \in (0, UB - R \cdot C/N)$. Thus,

$$\dot{W}(t) \leq -\frac{2(1 - \beta)}{1 + \beta} \cdot \frac{C \cdot K_p}{N} \cdot \left[\Delta W(q_0^* + \Delta q) + \frac{R \cdot C}{N}\Delta q\right] \tag{5.21}$$

for $t \in [t_1' + R, t_2]$.
On the other hand,

$$\int_{t_1'+R}^{t_2} \dot{W}(s)ds = W(t_2) - W(t_1' + R) \geq \frac{R \cdot C}{N} - UB. \tag{5.22}$$

It follows from (5.21) and (5.22) that,

$$\frac{R \cdot C}{N} - UB \leq -\frac{2(1 - \beta)}{1 + \beta} \cdot \frac{C \cdot K_p}{N} \cdot (t_2 - t_1' - R) \cdot \left[\Delta W(q_0^* + \Delta q) + \frac{R \cdot C}{N}\Delta q\right],$$

i.e.,

$$t_2 - t_1' - R \leq \frac{UB - \frac{R \cdot C}{N}}{\frac{2(1-\beta)}{1+\beta} \cdot \frac{C \cdot K_p}{N} \cdot \left[\frac{R \cdot C}{N}\Delta q + \Delta W(q_0^* + \Delta q)\right]}.$$

With the definition of T_1 in the theorem, we have $t_2 - t_1' \leq T_1 + R$. Therefore, it follows from (5.20) that

$$q(t) \leq \inf_{\substack{\Delta q > 0, \\ \Delta W \in [0, UB - R \cdot C/N]}} \left\{(q_0^* + \Delta q) + \left(\frac{N}{R} \cdot UB - C\right) \cdot (T_1 + R)\right\}, \tag{5.23}$$

i.e., $q(t) \leq UQ$ for $t \geq 0$, which indicates that UQ is the upper bound of the RED queue length. Since the packet loss in a RED queue is proportional to the queue length, the derived queue length upper bound also reflects the upper bound of packet loss rate.

We finally show that $LB_2 > 0$ is a lower bound of $W(t)$ for $t \geq t_1$, i.e., if $W(t) = LB_2$ at time $t = t_3 > t_1$, then $\dot{W}(t_3) \geq 0$.

Based on (5.17) and (5.23), we have

$$\dot{W}(t) \geq \frac{\alpha}{R} - \frac{2(1-\beta)}{1+\beta} \cdot \frac{UB^2}{R} \cdot K_p \cdot UQ \tag{5.24}$$

for $t \geq 0$, so that

$$\int_{t_3-R}^{t_3} \dot{W}(s)ds \geq \alpha - \frac{2(1-\beta)}{1+\beta} \cdot UB^2 \cdot K_p \cdot UQ$$

i.e.,

$$W(t_3 - R) \leq LB_2 + \frac{2(1-\beta)}{1+\beta} \cdot UB^2 \cdot K_p \cdot UQ - \alpha. \tag{5.25}$$

It follows from (5.23) and (5.25) that

$$\dot{W}(t_3) \geq \frac{\alpha}{R} - \frac{2(1-\beta)}{1+\beta} \cdot \frac{LB_2 \cdot UW}{R} \cdot K_p \cdot UQ$$

with

$$UW := LB_2 + \frac{2(1-\beta)}{1+\beta} \cdot UB^2 \cdot K_p \cdot UQ - \alpha.$$

Therefore, $\dot{W}(t_3) \geq 0$ if LB_2 is chosen to satisfy

$$LB_2 \cdot UW \cdot K_p \cdot UQ = \frac{\alpha(1+\beta)}{2(1-\beta)}, \tag{5.26}$$

and thus LB_2 is the lower bound of $W(t)$ for $t \geq t_1$.   □

5.3.3 Performance evaluation

In this subsection, analytical results obtained using Matlab and simulation results using NS-2 are given to validate the theorems and evaluate the system performance with different parameters.

AIMD parameter pairs

First, we investigate how the AIMD parameter pair (α, β) affects the bounds on the window size and queue length. Let N, R, C and K_p be constants: $N = 10$, $R = 0.1$ s, $C = 1000$ packets/s and $K_p = 0.01$. The AIMD(α, β) pairs are chosen to be TCP-friendly, varying from $(9/5, 1/4)$ to $(3/31, 15/16)$, and the results are given in

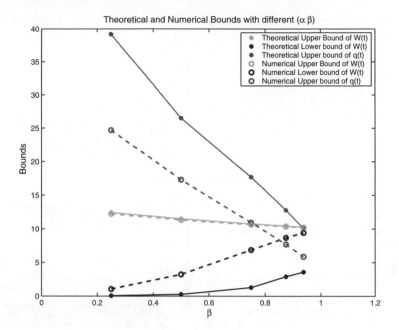

Figure 5.5 Theoretical and numerical bounds of the window size and queue length with different (α, β).

Figure 5.5. It can be seen that for the upper and lower bounds of the window size and the upper bound of the queue length, the numerical results are all within the bounds given by Theorem 5 and Theorem 7, which verifies the correctness of the theorems. In addition, the upper bound of the window size given by the theorem is very tight. The one for queue length is a loose bound as mentioned in the proof of Theorem 7. The theoretical lower bound of the window size is not tight because of the approximation of $\dot{W}(t)$ in (5.24). How to find a tight lower bound for window size is still an open issue.

In ideal cases, the window size should converge to $R \cdot C/N$, which is 10 packets per *rtt* in the above cases. The results in Figure 5.5 show that with a smaller value of α and a larger value of β, the AIMD flows have less oscillation amplitude around the optimal operation point, so they can utilize network resources more efficiently with less delay and loss in steady state. This is because, with a smaller value of α, the AIMD flows overshoot the available bandwidth in a slower pace; with a larger value of β, the AIMD flows will not decrease drastically for any single packet loss. Also, as shown in Figure 5.5, the upper bound of the queue length becomes smaller w.r.t. β; thus, the average queueing delay (and thus loss rate) becomes smaller in steady state.

Figure 5.6 shows the traces of TCP flows and those of AIMD(0.2, 0.875) flows. Here, the parameter values are $N = 10$, $C = 10^4$ packets/s, $R = 0.05$ s and $K_p =$

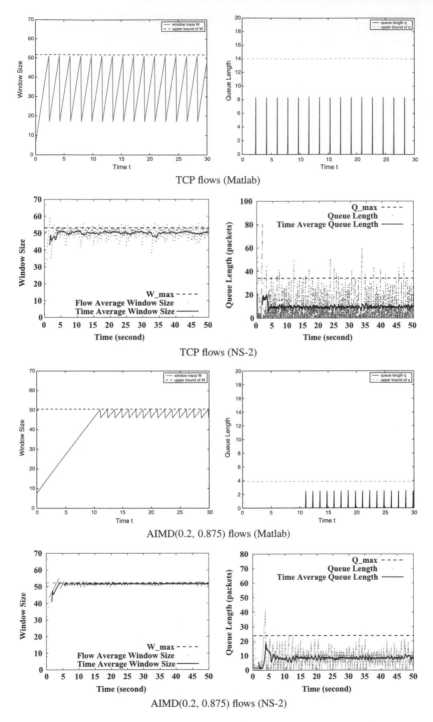

Figure 5.6 Traces of window size and queue length, $N = 10$, $C = 10\,000$ packets/s, $R = 0.05$ s and $K_p = 0.005$.

0.005. For NS-2 simulations, we set Q_{min} of the RED queue to 20 packets. Therefore, the upper bound of each flow's window size should be enlarged by $Q_{min}/N = 2$ packets, and the upper bound of the queue length should be enlarged by $Q_{min} = 20$ packets. Since the window size and queue length in the fluid-flow model are related to their ensemble averages in a real system, we compare the theoretical bounds with both the instantaneous average window size among all flows and its time average over a cycle. Both the computed results using Matlab and the simulation results using NS-2 show that, although the window variation of AIMD(0.2, 0.875) in steady state is smaller, it takes a longer time for AIMD(0.2, 0.875) flows to converge to steady state. Simulation results also demonstrate the tightness of the upper bound of the window size. Another interesting observation is that, although the upper bound of the queue length is not tight compared with the time average of the queue length, it is close to the maximum instantaneous queue length in steady state.

Considering that the traffic flows in future Internet might be a mixture of different types requiring the support of transport layer protocols such as AIMD with different parameters, we further study the performance of the AIMD/RED system with heterogeneous flows. Figure 5.7 shows the window trace and queue length when TCP flows and AIMD(0.2, 0.875) flows share the bottleneck. Parameters are firstly chosen as $C = 10^4$ packets/s, $K_p = 0.005$, and $R = 0.05$ s for five TCP flows competing with five AIMD(0.2, 0.875) flows. For comparison, we also choose $C = 2 \times 10^4$ packets/s, $K_p = 0.005$, and $R = 0.05$ s for ten TCP flows and ten AIMD(0.2, 0.875) flows. As shown in Figure 5.7, the TCP window size sustains larger oscillation than that of AIMD(0.2, 0.875), but their average window sizes are close to each other.

Note that the bounds for TCP or AIMD(0.2, 0.875) flows are only used for comparison purposes. The simulation results show that the time average window size of all flows are below the *UB* of TCP flows. This suggests that we can use the upper bounds of AIMD flows with the smallest value of β among the mixed traffic to estimate the upper bounds for systems handling heterogeneous traffic flows.

Theoretical bounds for a heterogeneous-flow AIMD/RED system can be obtained by applying a similar approach. But it is noted that the upper bound of the window size derived for the heterogeneous-flow system by this method is usually not as tight, because the interactions among heterogeneous flows are implicitly encapsulated in the fluid-flow model and not easy to calculate. The theoretical bounds of the AIMD/RED system with heterogeneous flows have been derived in [105].

Impact of system parameters

In the following, we study how the parameters N, R, C and K_p affect the bounds of the window size and queue length. We choose the (α, β) pair to be $(1, 0.5)$ and $(0.2, 0.875)$, and obtain the results with different network parameters shown in Tables 5.1 and 5.2.

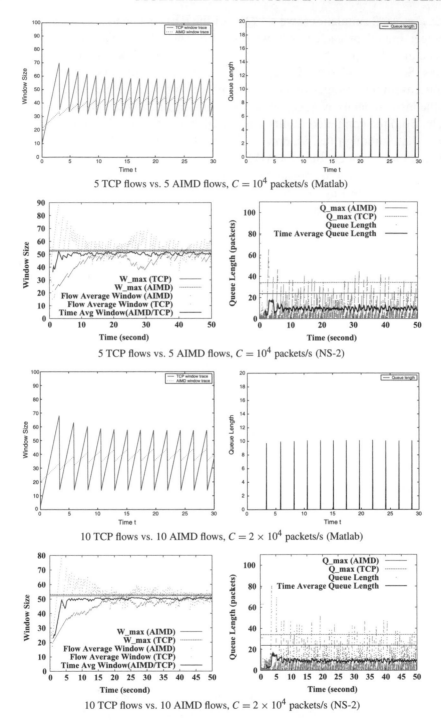

Figure 5.7 Heterogeneous flows, $K_p = 0.005$, and $R = 0.05$ s.

Table 5.1   AIMD/RED system bounds with $(\alpha, \beta) = (1, 0.5)$.

No.	N	R	C	K_p	(W^*, q^*)	UB Num	UB Ana	LB_2 Num	LB_2 Ana	UQ Num	UQ Ana
1	10	0.02	10^3	0.01	$(2, 37.5)$	4.04	4.41	1.52	0.09	51	147.5
2	10	0.05	10^3	0.01	$(5, 6)$	6.60	6.80	2.13	0.32	28	43.3
3	20	0.05	2E3	0.005	$(5, 12)$	6.60	6.80	2.12	0.38	56	78.0
4	10	0.05	10^3	0.005	$(5, 12)$	6.82	7.10	2.78	0.66	39	54.6
5	10	0.4	10^3	0.005	$(40, \frac{3}{16})$	41.30	42.02	14.14	0.11	10	23.2
6	10	0.05	10^4	0.005	$(50, 0.12)$	51.03	51.15	16.98	0.18	8	14.1
7	20	0.05	2E4	0.005	$(50, 0.12)$	51.00	51.20	8.91	0.068	15	23.1
8	100	0.05	10^4	0.005	$(5, 12)$	6.28	6.41	0.72	0.04	153	241.6
9	10^3	0.1	10^6	0.001	$(100, 0.15)$	101.0	101.02	0.026	2E-4	577	1024.2
10	10^4	0.1	10^6	0.001	$(10, 15)$	11.04	11.05	0.02	1.6E-4	6731	10785
11	10^4	0.1	10^6	0.005	$(10, 3)$	11.02	11.023	0.005	6.9E-6	5942	10349
12	10^4	0.1	10^6	0.01	$(10, 1.5)$	11.01	11.016	0.002	1.8E-6	5714	10248

Table 5.2   AIMD/RED system bounds with $(\alpha, \beta) = (0.2, 0.875)$.

No.	N	R	C	K_p	(W^*, q^*)	UB Num	UB Ana	LB_2 Num	LB_2 Ana	UQ Num	UQ Ana
1	10	0.02	10^3	0.01	$(2, 37.5)$	2.81	3.03	1.76	0.59	55	135.5
2	10	0.05	10^3	0.01	$(5, 6)$	5.50	5.63	4.19	1.77	18	31.2
3	20	0.05	2E3	0.005	$(5, 12)$	5.51	5.65	4.19	1.65	35	65.2
4	10	0.05	10^3	0.005	$(5, 12)$	5.62	5.80	4.27	2.10	29	48.7
5	10	0.4	10^3	0.005	$(40, \frac{3}{16})$	40.25	40.29	36.79	5.38	3	5.2
6	10	0.05	10^4	0.005	$(50, 0.12)$	50.23	50.26	45.93	6.31	2	3.8
7	20	0.05	2E4	0.005	$(50, 0.12)$	50.23	50.26	43.99	3.24	4	7.1
8	100	0.05	10^4	0.005	$(5, 12)$	5.34	5.46	3.76	1.39	67	83.8
9	10^3	0.1	10^6	0.001	$(100, 0.15)$	100.20	100.21	39.26	0.025	127	211.2
10	10^4	0.1	10^6	0.001	$(10, 15)$	10.22	10.23	2.02	0.02	1667	2361.4
11	10^4	0.1	10^6	0.005	$(10, 3)$	10.208	10.211	0.07	9.7E-4	1355	2158.8
12	10^4	0.1	10^6	0.01	$(10, 1.5)$	10.205	10.207	0.015	2.4E-4	1266	2111.5

Round-trip delay and link capacity   First, compare rows 1 and 2 in both tables. By enlarging the delay from 0.02 s to 0.05 s (by 2.5 times), the upper bound of window sizes only increases by 1.54 times and 1.86 times for TCP and AIMD(0.2, 0.875), respectively, which means a larger delay reduces the relative oscillation amplitude of the window size. In addition, the upper bound of the queue length is decreasing. A similar trend can be found by comparing rows 4 and 5 in both tables. This is a surprising result. From [98], a longer delay may drive the system from a stable to an unstable condition. This can be explained as follows. A larger delay means that the window size increasing speed (in terms of packets per second) during the additive increase period is smaller, and the AIMD flows will overshoot the network capacity at a slower pace; thus, the upper bound of the window size is closer to the optimal operating point, and the maximum queue length is smaller.

Similar results are found if we compare rows 4 and 6 in both tables. By enlarging the link capacity by ten times, the upper bound of window size is increased by 7.5 and 8.9 times, for TCP and AIMD(0.2, 0.875), respectively. Although enlarging the link capacity may drive the system from a stable to an unstable condition [98], the oscillating amplitude of the window size (relative to the equilibrium W^*) and queue length will actually decrease. The window and queue traces of ten TCP flows in a link with 10^3 packets/s and 10^4 packets/s are shown in Figure 5.8. The conclusion is that larger values of delay and link capacity will actually reduce the oscillating amplitude of the window size and queue length, and significantly reduce the maximum queueing delay.

Number of flows   Comparing rows 3 and 4, or rows 6 and 7 in Tables 5.1 and 5.2, we conclude that if we increase the number of flows and the link capacity proportionally, the bounds of window size are almost unaffected. With twice as many flows multiplexed in a link with twice the capacity, the upper bound of queue length increases less than twice. Therefore, the queueing delay bound is slightly reduced because of the multiplexing gain.

Comparing rows 6 and 8 in Tables 5.1 and 5.2, if we increase the number of flows in the same link, the $N \cdot UB$ becomes larger. In other words, the oscillation of the window size will increase significantly if the number of flows in a link increases, and the queueing delay will also increase significantly. This can be understood as N AIMD(α, β) flows will increase their windows by $N\alpha$ packets per *rtt*, and the larger the increasing rate during Additive Increase stage, the more significantly the flows will overshoot the link capacity. This suggests that we should either limit the number of TCP/AIMD connections in a link or use more conservative AIMD parameter pairs to ensure that the queueing delay (and also the loss rate) is less than a certain threshold.

RED parameter K_p   Comparing rows 2 and 4 in Tables 5.1 and 5.2, for a smaller value of K_p, the RED parameter will result in larger bounds of both window size and queue length.

The last four rows of Tables 5.1 and 5.2 are the upper bounds of the TCP/AIMD window size and queueing delay in a highly multiplexed, high bandwidth (tens of Gbps) and long delay (0.1 sec *rtt*) link. It can be seen that for TCP flows, the queueing delay can be bounded to 10.785 ms if the K_p is chosen to be 0.001. The delay bound can be slightly reduced to 10.349 ms and 10.248 ms if K_p is increased to 0.005 and 0.01, respectively. The results show that, although K_p can be adjusted to control the queueing delay in the system, the impact is limited for high bandwidth cases. Limiting the number of flows or using more conservative AIMD pairs are more

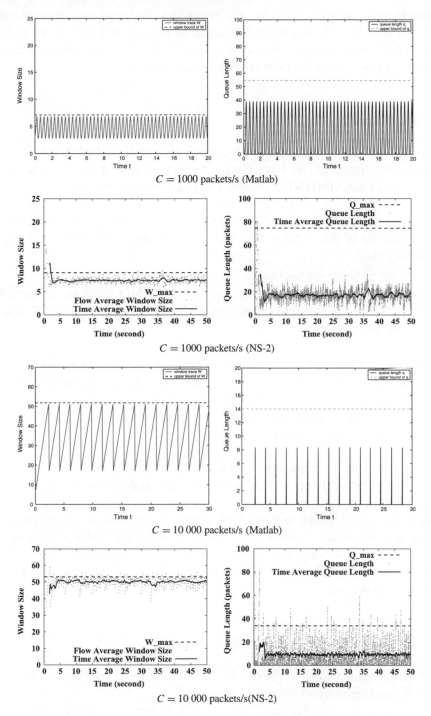

Figure 5.8 Bounds of TCP window size and queue length with different C.

effective in reducing queueing delay. For instance, if the number of flows is reduced to 100 or 1000, the queueing delay bound can be reduced to 0.241 ms or 1.079 ms, respectively. If using an AIMD parameter pair of (0.2, 0.875), the queueing delay for 10^4 flows with $K_p = 0.001$ can be bounded to 2.361 ms only.

5.4   Summary

In this chapter, we have first proved the asymptotic stability of AIMD/RED systems with delay-free marking. We have further derived bounds of the window size and queue length of the AIMD/RED system. The main conclusions are:

1. Larger values of delay and link capacity will actually reduce the amount of oscillation in the window size and queue length from their equilibrium in steady state.

2. If the link capacity and the number of TCP/AIMD flows are proportionally increased, the queueing delay will be slightly reduced so the multiplexing gain will increase slightly.

3. Although AIMD flows can adapt their sending rates according to the available bandwidth, a larger number of flows will lead to longer queueing delay in the AIMD/RED system. Thus, the number of AIMD connections in a link should be limited or more conservative AIMD parameters should be used to bound the queueing delay and loss.

The theorems stated in this chapter can also help to predict and control the system performance of future Internet with higher data rate links multiplexing more flows with different parameters.

There are many interesting research issues worth further investigation. For example, how to deploy effective admission control for TCP/AIMD flows to bound delay and loss, how to adapt the AIMD parameter pair to ensure that the system can converge to the equilibrium quickly enough and to control the queueing delay and loss in the network, etc.

5.5   Problems

1. Consider the fluid-flow model of the AIMD/RED system and answer the following questions:

$$\frac{dW(t)}{dt} = \frac{\alpha}{R} - \frac{2(1-\beta)}{1+\beta} W(t) \frac{W(t)}{R} K_p q(t),$$

$$
\frac{dq(t)}{dt} =
\begin{cases}
\dfrac{N \cdot W(t)}{R} - C, & q > 0 \\[2ex]
\left\{ \dfrac{N \cdot W(t)}{R} - C \right\}^{+}, & q = 0,
\end{cases}
$$

where C is the link capacity, N is the number of AIMD flows sharing the link, R is the round-trip time of each flow, α and β are the increase rate and decrease ratio of the AIMD flows, $W(t)$ is the window size of a flow at time t, $q(t)$ is the queue length at time t, and K_p is the RED parameter.

(a) Derive the equilibrium point (W_0^*, q_0^*) of the system.

(b) What is the link utilization at equilibrium?

(c) For $N = 1$, at equilibrium, what is the average packet loss rate?

2. Prove the following theorem: *The equilibrium point of (5.3) is asymptotically stable for all $K_p > 0$.*

3. Consider an Additive Increase and Additive Decrease (AIAD) congestion controller: if there is no packet loss, the congestion window size is increased by a packets; otherwise, it is decreased by b packets. Let one AIAD flow occupy a link with RED queue management. Assume the round-trip time of each flow is a constant, $R(t) = R$.

(a) Use the fluid-flow model to describe the AIAD/RED system.

(b) Derive the equilibrium point of the above system.

(c) Ignoring the feedback delay, will the fluid-flow model of the AIAD/RED system be asymptotically stable?

(d) Considering the feedback delay, derive the upper bound of the congestion window size.

4. Let $N = 100$ TCP flows share a link with capacity $C = 100\,000$ packets/s. The round-trip time (without queueing delay) of each flow is $R = 50$ ms, and the RED queue parameter is $K_p = 0.001$.

(a) Derive the upper bound of the congestion window size.

(b) What is the upper bound of packet loss rate?

(c) Use NS-2 to obtain the congestion window size and packet loss rate. Compare the simulation results with the bounds derived.

5. Consider the multiple-bottleneck network shown in the figure below. Three groups of TCP flows share the two links $r_0 r_1$ and $r_1 r_2$. The number of flows and round-trip time (without queueing delay) of the ith group are N_i and R_i, respectively. All routers are RED-capable with the RED parameter $K_p (> 0)$.

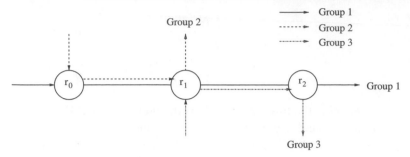

(a) For the multiple-bottleneck system, build a mathematical model to investigate the ensemble averages of the flow congestion window sizes and queue lengths at r_0 and r_1.

(b) Ignoring feedback delay, what are the ensemble averages of window sizes and queue lengths in equilibrium?

(c) Ignoring feedback delay, is the multiple-bottleneck system asymptotically stable?

6. Let five TCP flows compete with five AIMD(0.2, 0.875) flows in a link with capacity 10^4 packets/s. The router is RED-capable with the RED parameter $K_p(= 0.005)$. The round-trip times without queueing delay for both flows are 0.05 s. Consider the feedback delay and answer the following questions.

(a) Derive the upper bounds of the flows' window sizes.

(b) Derive the upper bounds of the queue length.

(c) We double the link capacity to 20 000 packets/s, and let ten TCP flows compete with ten AIMD(0.2, 0.875) flows in the link. K_p and the flows' round-trip time without queueing delay are still 0.005 and 0.05 s, respectively. Derive the upper bounds of the flows' window sizes and the upper bounds of the queue length. Compare the results with the answers in (a) and (b).

6

AIMD in Wireless Internet

6.1 Introduction

Wireless systems and the Internet are anticipated to converge into a ubiquitous information transport infrastructure, allowing users to access the IP-based hybrid wireless and wireline networks for multimedia services anywhere, anytime. However, there are great challenges in providing *timely* delivery services with quality of service satisfaction for delay-sensitive multimedia applications in wireless Internet.

First, wireless channels exhibit inherent severe impairments. For a given wireless channel, with FEC coding, the residual transmission error rate may still be non-negligible and visible to the upper layer protocols. In addition, the instantaneous error rate fluctuates over time. To reduce the transmission errors visible to the upper layers, the link layer can incorporate Automatic Repeat reQuest (ARQ) to detect transmission errors and request retransmissions locally [106, 107]. For delay-sensitive applications, it is necessary to terminate retransmission after a prescribed number of attempts. With ARQ, the wireless link throughput (the number of frames successfully transmitted over the link per unit time) and delay are random. Also, channel state dependent (CSD) transmission and scheduling schemes can be introduced to achieve multi-user diversity gain and to improve link utilization, which also introduce throughput and delay variation over the wireless links [108, 109]. Recent wideband wireless communications also deploy adaptive modulation and coding (AMC), so the instantaneous bit rate of the wireless link is adaptive to the channel condition. The variation in link throughput and delay is still visible to the upper layers. Hence end-to-end delay control faces new challenges [110].

Second, emerging Internet-based multimedia applications can use scalable and error-resilient source coding to adapt to network dynamics. However, they require timely delivery services. Specifically, before a packet is sent, the application

Multimedia Services in Wireless Internet: Modeling and Analysis   Lin Cai, Xuemin Shen and Jon W. Mark
© 2009 John Wiley & Sons, Ltd

determines the most appropriate information in that packet.[1] It is up to the transport layer protocol to provide a timely delivery service, i.e., to statistically guarantee the end-to-end delay of the packet, where end-to-end delay measures the time a packet traverses the network in the transport layer.

In this context, a key challenge is how transport layer protocols can guarantee end-to-end delay in wireless Internet. To meet the challenge, the end-to-end delay performance of cross-domain congestion-controlled flows should be obtained. We focus on window-based transport layer protocols in this chapter. Window-based congestion control mechanisms are simple to implement, and the *acknowledgment self-clocking* property is desired for delay control over time-varying wireless links.

In the literature, there are two approaches to delay analysis for TCP over wireless links with link level ARQ. The first is a Markov process analysis, with the assumption that the arrival process of TCP packets at the wireless links follows a Bernoulli, Poisson, or on/off process [111–114]. However, for a closed-loop window-controlled flow, new packets are injected into the network when *ack*s of previous packets are received by the sender, so the arrival process is highly correlated. Therefore, the Markov assumption for a window-controlled flow is impractical in general [115].

The second approach uses fluid models. In [100, 116, 117], the queue-length process and the rate at which a source transmits data are approximated by fluid models. However, a fluid model is only suitable in an environment where the insertion delay is short compared with the total transfer time and does not vary significantly from packet to packet. Thus, the fluid model approach is not suitable for delay analysis in wireless links with ARQ.

Different from the previous work, here we introduce a novel analytical framework for quantifying the QoS performance of window-controlled flows in hybrid networks [118]. *Token* control is used to emulate the window-based flow and congestion control. With the token emulation, the queueing delay distribution and packet loss rate of cross-domain flows have been obtained as a function of the sender window size and the wireless link throughput distribution.

The token-emulation analysis captures the traffic characteristics of closed-loop window-controlled flows, and it is applicable to various wireless links with different channel characteristics and packet transmission schemes. The analytical results reveal how the window size is related to delay distribution, which offers much-needed insights for tuning transport layer protocol parameters to provide timely delivery services. Since end-to-end delay is mainly related to the sender window size during the past *rtt*, and congestion control dynamically adjusts the window size in different *rtt*s, delay control and congestion control are of different timescales and can be studied separately. Thus, the analytical framework can be applied to

---

[1]Link layer fragmentation is not considered here, since delay-sensitive multimedia traffic usually has small packet size, and transport layer protocols can negotiate for the maximum segment size at the time of connection establishment. In the sequel, the term *packet* is used generically to represent the link *frame*, network *packet* and transport *segment*.

different window-based transport layer protocols using different congestion control algorithms.

The remainder of this chapter is organized as follows. Work available in the literature related to the material presented in this chapter is discussed in Section 6.2. Section 6.3 presents the system model and calculates the wireless link throughput distributions. In Section 6.4, the delay distribution and packet loss rate of window-controlled flows over hybrid networks are analytically obtained, and a delay control scheme is introduced. Based on the QoS performance analysis, we demonstrate how to appropriately select the protocol parameters in Section 6.5. Simulation results are given in Section 6.6. Section 6.7 summarizes the key features presented in this chapter.

6.2 Related Work

6.2.1 TCP over wireless networks

TCP was developed as a transport layer protocol to deal with information loss due to traffic congestion in wireline networks. The more hostile wireless channel presents a new set of issues that are not captured in the design of conventional TCP. The mainstream TCP variants, TCP Tahoe, TCP Reno, TCP New Reno and TCP-SACK, are all based on the AIMD congestion control mechanism, which uses packet losses to generate congestion indicators on the assumption that packet losses are mainly due to network congestion. However, this assumption may not hold in the wireless domain, which has a noticeable transmission error rate. Efforts have been taken in both the link layer and the transport layer to improve TCP performance over wireless links.

Link layer approaches try to reduce the transmission errors visible to TCP [19–21, 119]. FEC coding is used to enhance the error correction ability by introducing more redundancy. Since wireless channels usually introduce burst errors, FEC coding alone cannot efficiently correct them. In general, for a given wireless channel and certain FEC coding schemes, the residual transmission errors at the output of the decoder are taken care of by the link-level ARQ. However, the link-level ARQ introduces more delay variation, which should be considered when designing a transport layer protocol for delay-sensitive applications. The work reported in this chapter serves this purpose.

Besides the link layer approaches, many schemes have been proposed to adjust the TCP behavior over wireless links [120–124], either with or without the assistance of the interface node, i.e., the base station (BS). With Explicit Congestion Notification/Explicit Loss Notification (ECN/ELN) [120, 121], the BS explicitly notifies the congestion or transmission losses to the TCP sender. Since ECN/ELN cannot recover transmission errors, either the link-level ARQ or a transport layer error recovery scheme is required when the application cannot tolerate excessive packet losses. For delay-sensitive applications, the link-level ARQ is preferable since it induces less delay and delay jitter.

I-TCP is a split connection approach [123]: the BS establishes a standard TCP connection with the correspondent host (CH) and a wireless TCP connection with the mobile host (MH), so the controls are conducted and optimized over different domains. However, split connection approaches have significant overhead during handoff, and the end-to-end delay and delay jitter are difficult to control with two connections. With snoop TCP [122, 125], the BS snoops the data packets and *acks*, and buffers the unacknowledged data. Once a packet loss due to transmission error is detected, the buffered packet is retransmitted locally. This approach is similar to the link-level ARQ, except that the local retransmissions are performed in the transport layer. Since it may take a longer time for the transport layer to detect transmission errors, for delay-sensitive applications it is preferable to use the link-level ARQ to recover transmission errors.

With M-TCP [124], the BS detects disconnections in the wireless domain and instructs the sender to freeze its timer and window. Once the connection is regained, the sender is notified that it can send at the same speed as before disconnection. Similarly, the MH can detect disconnections and instruct the sender to freeze its timer and window, as proposed in [126, 127]. The window freeze schemes are useful when temporary disconnection occurs due to a deep fading channel condition or during the handoff period; these schemes address issues different from those dealt with in this chapter, and can be deployed in the AIMD protocol described in this chapter to deal with temporary disconnections.

Readers interested in this topic can check Dr Jianping Pan's annotated pointers at [128].

6.2.2   Using *rwnd* to enhance TCP performance

Using *rwnd* to enhance TCP performance has been proposed in the literature [80, 129–131]. It is proposed in [129] to auto-tune *rwnd* to improve the throughput; in [80], *rwnd* is used to enhance fairness and reduce packet losses in the wired link. In the approach discussed in the following sections, the optimal *rwnd* is set not only to efficiently utilize the time-varying wireless link and avoid buffer overflow at the BS, but also to bound the delay outage rate, i.e., the percentage of packets with end-to-end delay exceeding the pre-defined threshold. In [130] and [131], *rwnd* is used to enhance TCP performance for hybrid IP and third-generation wireless/wireline networks, by adaptively adjusting *rwnd* according to the queue length (or free buffer size) at the interface node. Since the *rwnd* is determined on the assumption that there is no delay and loss in the wireline domain, the performance degrades when the *rtt* is longer than 100 ms or the packet loss rate due to congestion in the wireline domain exceeds 0.1%. Instead of frequently changing *rwnd* according to the current queue length, in the following, the optimal *rwnd* is obtained with consideration of the link

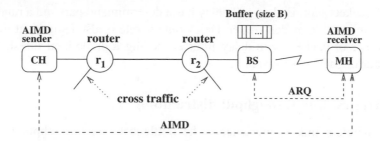

Figure 6.1 System model [118]. Reproduced by permission of ©2007 IEEE.

profile, flow *rtt* and application QoS requirements; the simulation results show the robustness of this approach.

6.3   System Model

As shown in Figure 6.1, we consider the transmission of multimedia traffic from a CH to an MH, through a last-hop wireless link between a BS and the MH. The method used is also applicable to analyzing the performance of the reverse direction transmission.

In the wireless domain, a number of wireless channels with fixed raw capacity are assigned to a number of multimedia flows according to their QoS requirements, since wireless resources are channelized in cellular systems. When new and handoff calls arrive, or existing calls depart or terminate, the allocated resources are adjusted only infrequently. So long as the delay bound for delay-sensitive applications is not violated, the low-persistent ARQ scheme can be deployed in wireless links so that a corrupted packet will be retransmitted *immediately* after the current transmission [107]. (In some situations, there will be a non-zero retransmission delay, which will be discussed separately in Subsection 6.4.4.)

6.3.1   QoS indexes for delay-sensitive applications

For delay-sensitive multimedia applications over best-effort and highly dynamic networks, scalable source coding schemes such as MD coding are anticipated to be widely deployed [13].

As discussed in Section 6.1, with the MD coding technique, the source encoder determines an optimal bit-rate, according to the current sending rate in the transport layer, and the maximum tolerable packet loss rate. Also, for delay-sensitive applications, packets that have suffered excessive delay are useless and will be discarded by the receiver. Therefore, besides flow throughput and packet loss rate, an important QoS index is delay outage rate, which is the ratio of the number of packets with end-to-end delay exceeding a threshold predefined by the application to the total number of packets sent. End-to-end delay has a deterministic part and a random part; the latter is referred to as delay jitter. Delay outage rate is also equivalent to the ratio of the number of packets with delay jitter exceeding a predefined threshold to the total number of packets received.

6.3.2 Wireless link throughput distribution

Since the wireless link is presumably the bottleneck, its throughput distribution is closely related to end-to-end delay. Link throughput depends on the channel characteristics, and the transmission and error control schemes used in the physical and link layers. In this text, the packet-level channel characteristics observed at the link layer are collectively referred to as the channel profile. Given the channel profile and the link layer packet transmission scheme, the link throughput distribution can be obtained. The notations used in this chapter are listed in Table 6.1 for easy reference.

Throughput distribution with persistent transmission

Finding an efficient packet transmission scheme for information transfer over the wireless Internet is an active research topic [108, 109, 132, 133]. As discussed in Chapter 2, the wireless channel can be modeled as a finite-state Markov chain; the value of λ and the memory of the Markov chain highly influence the optimality of a transmission scheme. In general, if the value of λ is small, a traditional *persistent* transmission scheme is very efficient. If the channel has a large λ, the CSD transmission scheme is more efficient [108]; at the end of the transmission, the transmission may be suspended for some time slots if the channel quality during the past time slot is poor. The duration of the transmission suspension is determined according to λ and the cost of failed transmission [108].

The duration of a slot equals the transmission time of a packet. With a *persistent* transmission scheme, one packet is transmitted per slot no matter what the previous channel condition was. The probability of successfully transmitting x packets in n slots is

$$T_n(x) = \binom{n}{x} \sum_{b_1,\dots,b_M} \prod_{i=1}^{M} (\pi_i(1 - e_i))^{b_i} \sum_{a_1,\dots,a_M} \prod_{i=1}^{M} (\pi_i e_i)^{a_i}, \qquad (6.1)$$

where $b_1 + \cdots + b_M = x$, $a_1 + \cdots + a_M = n - x$, $0 \le b_i \le x$, and $0 \le a_i \le n - x$.

The average link throughput is $(1 - p_e)$ packets per slot, independent of λ. However, λ plays an important role in link throughput distribution. This can be illustrated by a simple example. Consider two channels that are modeled by two-state Markov chains with the parameters listed in Table 6.2. Their two-slot link throughput distributions are shown in Figure 6.2. Although these two channels have the same steady state distributions, packet error rates, and average link throughputs, their link throughput distributions are different due to different values of λ. Generally speaking, with higher user mobility, λ is smaller, and the link throughput distribution is higher

Table 6.1   Notation for Chapter 6.

Symbol	Description
γ	Received instantaneous signal-to-noise ratio
g_i	Channel state defined by level i and level $i+1$
π_i	Steady-state probability of state g_i
e_i	Packet error rate of state g_i
p_e	Average packet error rate over the wireless link
$P_{i,j}$	State transition probability from state g_i to state g_j
$N(\Gamma_i)$	Level crossing rate at Γ_i
μ	Doppler frequency shifts
λ	*Memory* of the Markov chain
$T_n(x)$	Probability of x successful transmissions in n slots over the wireless link
t_s	Time to transmit a packet over the wireless link
W	Sender window size
R	Round-trip time without queueing and retransmission delay in the wireless link
D	The maximum tolerable delay jitter
B	BS buffer size
q	BS queue length
d_q	BS queueing delay
L	Probability of packet loss due to BS buffer overflow
$\vec{W}$	Vector $[W_1\ W_2\ \cdots\ W_N]$
$T_n(x; P_i)$	Probability of successfully transmitting x packets from the ith flow in n slots, and P_i equals the share of bandwidth assigned to the ith flow
S	Sum of windows of all flows sharing the wireless link
d_l	Local retransmission delay
r_t	Number of link-layer local retransmissions
$\tilde{d}_q$	Delay in the new arrival queue

in the middle (e.g., $T_n(n/2)$); otherwise, the distribution is higher at the two ends (e.g., $T_n(0)$, $T_n(n)$).

When λ approaches zero, T_n becomes a binomial random variable with the following distribution:

$$T_n(x) = \binom{n}{x} \left(\sum_{i=1}^{M} \pi_i(1 - e_i) \right)^x \left(\sum_{i=1}^{M} (\pi_i e_i) \right)^{n-x} = \binom{n}{x} (1 - p_e)^x (p_e)^{n-x}. \quad (6.2)$$

Table 6.2   Parameters of two two-state Markov chains.

	π_1	π_2	e_1	e_2	p_e	$P_{1,2}$	$P_{2,1}$	λ
Case (a)	0.1	0.9	1	0	0.1	0.9	0.1	0
Case (b)	0.1	0.9	1	0	0.1	0.5	0.055556	0.44444

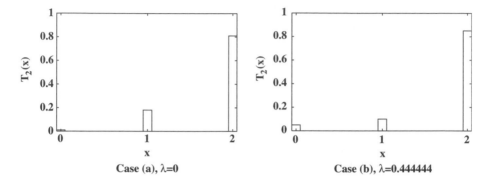

Figure 6.2 $T_2(x)$ of two two-state Markov channels.

Throughput distribution with CSD transmission scheme

With a CSD transmission scheme, to calculate the throughput distribution in n slots, the probability and throughput of each possible state trajectory, $\{G_1, G_2, \ldots, G_n\}$, should be calculated. $T_n(x)$ is the sum of the probabilities of those trajectories with throughput equal to x packets. The number of trajectories increases exponentially with n, and so does the computational complexity. To simplify the computation, we can further explore the convergent speed of the Markovian channel to obtain the approximation of the throughput distribution.

Equation (2.7) shows that $\Pr\{G_n|G_1\}$ will exponentially converge to π. If $|\Pr\{G_k|G_1\} - \pi|$ is smaller than a small positive value ϵ, we can assume that the channel states after the kth slot are independent of G_1. Thus, n slots can be divided into n/k blocks, and each block has k consecutive slots. The number of packets being successfully transmitted in n slots, T_n, is approximately the sum of n/k i.i.d. random variables T_k (the number of successful transmissions in k slots). Therefore, the probability distribution of T_n can be approximated as the convolution of the distribution of T_k for $n/k - 1$ times, where k is the minimum integer satisfying $|\Pr\{G_k = i|G_1\} - \pi_i| \le \epsilon$. According to (2.7), k is proportional to $\log_\lambda \epsilon$. When n is much larger than k, T_n can be further simplified as a Gaussian random variable with mean $(n/k)E[T_k]$ and variance $(n/k)\text{Var}(T_k)$; in this way, the computational complexity of the throughput distribution is only $O(M^k)$.

6.4   Analytical Model for Window-controlled Flows

Window dynamics for congestion control have been studied in Chapter 4. To maintain network stability and achieve fairness among coexisting flows, window-based transport layer protocols adjust *cwnd* according to network conditions. On the other hand, end-to-end delay is mainly related to the window size during the past *rtt*. Since congestion control and delay control are of different timescales, they can be studied separately. In this chapter, we focus on the delay performance with a fixed window size, i.e., the number of in-flight packets without *ack* is fixed. Then, we substantiate the analysis by developing a delay control scheme for window-controlled flows.

Transmission time is partitioned into slots, each long enough to transmit one packet over the wireless link. To facilitate the ensuing discussions in this section, we make the following assumptions; the impacts of relaxing these assumptions will be discussed in Subsection 6.4.6.

Assumptions:

1. All packets of a given flow have the same packet size and travel through the same route.

2. Let the *rtt* without queueing and retransmission delay in the wireless link be R slots. R can be assumed approximately constant; since the wired links exhibit less fluctuation than the wireless ones, queueing delay and delay variation is more dominant in the wireless domain and can absorb delay variation in the wireline domain.

3. The receiver acknowledges each received packet without delay, and all *ack*s can be successfully transmitted without loss.

4. The sender always has data to send, i.e., it is a saturated sender.

5. The sender sets end-to-end round-trip timers conservatively, so there is sufficient time for the low-persistence ARQ scheme to recover from transmission errors. This assumption tends to hold, since the timeout value is estimated as the average of measured *rtt*s plus a conservative factor proportional to the measured standard deviation of *rtt*. Furthermore, coarse-grained timers with a granularity of 500 ms are implemented in practice.

6. The probability of packet losses due to failed retransmissions by the link-level low-persistence ARQ is negligible.

6.4.1   Single flow, sufficient buffer

We first consider one window-controlled flow with a fixed sender window size, W, occupying a wireless link. Let the interface buffer size B be no less than W, so that

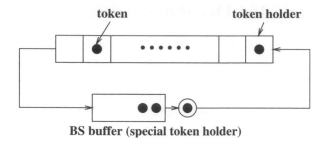

Figure 6.3 Token emulation, single flow [118]. Reproduced by permission of ©2007 IEEE.

the BS buffer will not overflow. After sending out a window of packets, the sender can resume sending only when it receives a new *ack*.

Given a window size, we have the knowledge of the number of in-flight packets plus *ack*s, but we cannot tell how many of them are data packets or *ack*s. Also, without packet loss and re-ordering, when the *ack* for a packet with sequence number x returns to the sender, the packet with sequence number $x + W$ can be sent out. To capture the relationship between *ack*s and data packets, we use *token* control to emulate the window control. A packet, which can be data or *ack*, needs a token to traverse the network, so that the sender should have enough tokens for outgoing packets. When the receiver receives a data packet with an accompanied token, an *ack* is returned with the token. The sender generates and adjusts the total number of tokens in-flight according to the flow and congestion control mechanisms, i.e., the number of tokens in-flight is always the same as the sender window size. Note that the term *token* is a virtual entity, which is used for easy presentation and analysis purposes.

When the sender window size is W, this is equivalent to W tokens circulating along the path. It takes R slots for a token to finish a round trip if there are no queueing delay and retransmissions at the BS. The round-trip path can be represented by R token holders. A token shifts to the next holder along the path per time slot, unless it is in the BS buffer, which is a special token holder. If a packet at the front of the BS queue is transmitted successfully, the token with that packet can shift from the BS buffer to the next token holder,[2] as shown in Figure 6.3. The number of tokens in the $R - 1$ non-special token holders equals the number of successful transmissions over the wireless link during the past $R - 1$ slots; the remaining tokens are queued in the BS buffer, the special token holder.

Proposition 3 *For a flow with window size W occupying a wireless link with sufficient buffer ($B \geq W$), if during the past $R - 1$ slots, n packets were transmitted*

---

[2]Token holders act similar to feedback shift registers.

successfully over the wireless link, the BS queue length, q, can be determined by W and n : q = (W − n)⁺, where 0 ≤ n ≤ R − 1 and (x)⁺ = max(x, 0).

To fully utilize the wireless resources, i.e., the wireless link should not be idle whenever the link layer decides to transmit, q should be greater than zero all the time. Therefore, the sufficient condition to fully utilize the wireless link is $W \geq R$. From Proposition 3, when a packet arrives at the BS buffer, the probability of the queue length equal to x packets is the probability that $W - x$ packets were transmitted successfully over the wireless link during the past $R - 1$ slots, denoted as $T_{R-1}(W - x)$, which has been derived in Subsection 6.3.2.

The time duration from the instant a token leaves the BS buffer to the instant it leaves the BS buffer again, $R + d_q$, equals the time that W packets are successfully transmitted over the wireless link, where d_q is the queueing delay in the BS buffer. Therefore, the probability of $R + d_q > R + D$ equals the probability of fewer than W successful transmissions over the wireless link in $R + D$ slots.

Proposition 4 *For a flow with window size W (W ≥ R) occupying a wireless link with buffer size B (B ≥ W), the probability of queueing delay, d_q, exceeding a threshold D equals the probability of fewer than W successful transmissions over the wireless link in R + D slots:*

$$\Pr\{d_q > D | W\} = \sum_{x=0}^{W-1} T_{R+D}(x). \tag{6.3}$$

In Subsection 6.4.5, we will demonstrate how to apply Proposition 4 to statistically bound the delay by appropriately bounding the window size.

When $B \geq W \geq R$, there are no packet losses due to BS buffer overflow, and the flow throughput equals the average link throughput, e.g., it equals $1 - p_e$ packet per slot with the *persistent* transmission scheme.

6.4.2 Single flow, limited buffer

From Proposition 3, in the worst-case scenario, the maximum queue length is W. When $B < W$, buffer overflow may occur. With a limited buffer B ($B < W$) and a window size W ($W \geq R$), the probability of buffer overflow, $L(B, W)$, equals the probability of fewer than $W - B$ successful transmissions over the wireless link in the past $R - 1$ slots:

$$L(B, W) = \Pr\{q = B | W; B\} = \sum_{x=0}^{W-B-1} T_{R-1}(x). \tag{6.4}$$

When $W \geq R$, the minimum queue length is $W - R + 1$, and the queue length distribution is

$$\Pr\{q = x | W; B\} = T_{R-1}(W - x), \tag{6.5}$$

where $W - R + 1 \leq x < B$.

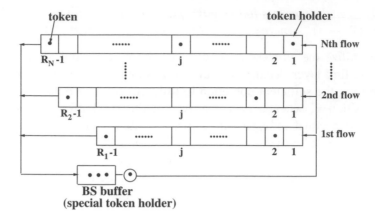

Figure 6.4 Token emulation, multiple flows [118]. Reproduced by permission of ©2007 IEEE.

When the queue length equals x, the probability of queueing delay exceeding D slots is $\sum_{i=0}^{x-1} T_{D+1}(i)$. The queueing delay distribution is given by

$$\Pr\{d_q > D | W; B\} = \sum_{x=W-R+1}^{B} \sum_{i=0}^{x-1} T_{D+1}(i) \cdot T_{R-1}(W-x). \qquad (6.6)$$

Equations (6.4) and (6.6) reveal the conflicting requirements in choosing the value of buffer size: a larger buffer may reduce the probability of buffer overflow at the cost of a larger probability of delay outage. To choose a proper buffer size, the cost of transmitting a packet suffering excessive delay and the cost of packet loss due to buffer overflow should be considered jointly to achieve the best tradeoff.

6.4.3 Multiple flows

When several window-controlled flows share a wireless link to achieve statistical multiplexing gains, the token emulation is still applicable.

Figure 6.4 shows N flows sharing a wireless link. For the ith flow, the minimal *rtt* and sender window size are denoted as R_i and W_i, respectively. Without loss of generality, let $R_1 \leq R_2 \leq \cdots \leq R_N$. Let P_i be the probability of transmitting packets from the ith flow, where $1 \leq i \leq N$. The mean number of the ith flow's tokens in the BS buffer equals $W_i - E[T_{R_i-1}]P_i$, where $E[T_{R_i-1}]$ is the mean number of successful transmissions over the wireless link in $R_i - 1$ slots. On the other hand, P_i equals the ratio of the mean number of the ith flow's tokens in the buffer to the total outstanding window size, i.e.,

$$P_i = \frac{W_i - E[T_{R_i-1}]P_i}{S - \sum_{j=1}^{N} E[T_{R_j-1}]P_j}, \qquad (6.7)$$

where $S = \sum_{i=1}^{N} W_i$. In other words,

$$\frac{W_i}{W_j} = \frac{W_i - E[T_{R_i-1}]P_i}{W_j - E[T_{R_j-1}]P_j}, \quad \text{for } 1 \leq i, j \leq N. \tag{6.8}$$

From (6.8), to assign the ith flow a P_i portion of the wireless bandwidth, W_i should satisfy the following condition:

$$W_i = \frac{P_i W_j}{P_j} + (E[T_{R_i-1}] - E[T_{R_j-1}])P_i, \quad \text{for } 1 \leq i, j \leq N. \tag{6.9}$$

Equation (6.9) gives the *necessary* and *sufficient* condition of proportional fairness for window-controlled flows. According to (6.9), for $P_i = 1/N$, the flows' window sizes are not proportional to their *rtt*s. To achieve fairness for flows with different *rtt*s, the number of coexisting flows and their *rtt*s should be taken into consideration. This reveals why the AIMD mechanism alone cannot achieve fairness for flows with different *rtt*s, no matter how the values of the (α, β) pair are chosen (such bias has been reported in the literature [134]). In Section 6.6, the simulation results validate (6.9).

Let $T_n(x; P_i)$ be the probability of successfully transmitting x packets of the ith flow in n slots. $T_n(x; P_i)$ can be calculated as

$$T_n(x; P_i) = \sum_{k=0}^{n-x} \binom{x+k}{x} P_i^x (1 - P_i)^k T_n(x+k), \tag{6.10}$$

where $0 < P_i < 1$. Obviously, $T_n(x) = T_n(x; 1)$ and $T_n(x; P_i) + T_n(x; P_j) = T_n(x; P_i + P_j)$.

As shown in Figure 6.4, the path of each flow corresponds to a set of token holders. The jth token holder of the ith flow holds one token if a packet of the ith flow was successfully transmitted over the wireless link j slots earlier. The sum of tokens in the jth token holders of all paths can be either zero, if no packet is transmitted successfully over the wireless link j slots ago, or 1 otherwise. Thus, the maximum number of in-flight tokens outside the interface buffer is $R_N - 1$. To fully utilize the wireless resources, a sufficient condition is $S \geq R_N$.

Sufficient buffer   Let $B \geq S \geq R_N$. The number of the ith flow's tokens outside the BS buffer equals the number of ith flow packets successfully transmitted during the past $R_i - 1$ slots. Thus, the queue length distribution is

$$\Pr\{q = x | \vec{W}\}$$

$$= \sum \cdots \sum_{y_1 + \cdots + y_N = S-x} [T_{R_1-1}(y_1; 1) T_{R_2-R_1}(y_2; 1 - P_1) \cdots T_{R_N-R_{N-1}}(y_N; P_N)], \tag{6.11}$$

where $\vec{W}$ represents the vector $[W_1 \ W_2 \cdots W_N]$, and $S - R_N + 1 \leq x \leq S$.

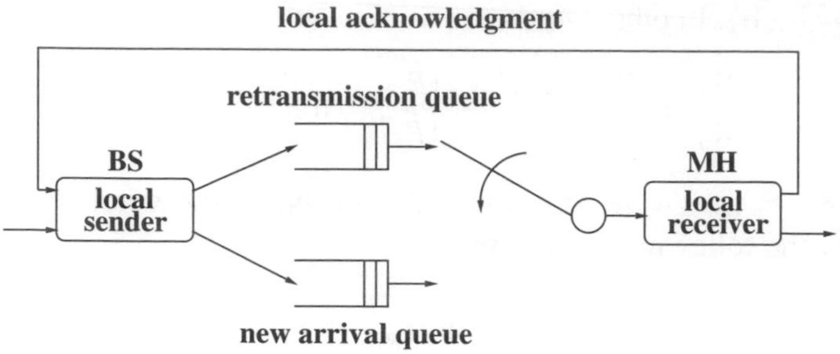

local acknowledgment

Figure 6.5 Priority queueing system.

The probability of queueing delay exceeding D slots is given by

$$\Pr\{d_q > D | \vec{W}\} = \sum_{x=S-R_N+1}^{S} \Pr\{q = x | \vec{W}\} \sum_{j=0}^{x-1} T_{D+1}(j), \qquad (6.12)$$

where $S \geq R_N$.

Limited buffer   When $B < S$, the probability of buffer overflow, $L(B, \vec{W})$, is

$$L(B, \vec{W}) = \sum_{x=B+1}^{S} \Pr\{q = x | \vec{W}\}. \qquad (6.13)$$

The probability of queueing delay exceeding D slots is given by

$$\Pr\{d_q > D | \vec{W}; B\} = \sum_{x=S-R_N+1}^{B} \Pr\{q = x | \vec{W}\} \sum_{j=0}^{x-1} T_{D+1}(j), \qquad (6.14)$$

where $S \geq R_N$.

6.4.4   Local retransmission delay

In Subsections 6.4.1–6.4.3, we assumed that local acknowledgment is received instantaneously after transmission, and local retransmission is triggered immediately afterward. In some situations, there may be a non-negligible delay between a packet being transmitted and its retransmission being triggered. At the BS, to reduce delay jitter, a higher priority is given to the retransmitted packets. Consider the non-preemptive priority queueing discipline shown in Figure 6.5. In this case, the delay jitter contains two parts: the queueing delay (in both the new arrival queue and the retransmission queue) and the local retransmission delay. This subsection analyzes the effect of local retransmission delay on the end-to-end delay distribution.

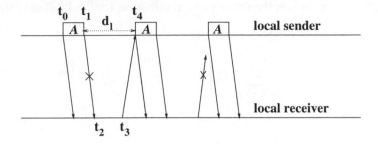

Figure 6.6 Local retransmission delay [118]. Reproduced by permission of ©2007 IEEE.

Consider the signal flow of local retransmission shown in Figure 6.6. Packet A is transmitted over the wireless link from time t_0 to t_1. At t_3, the local receiver sends a local negative acknowledgment back to the local sender. At t_4, since the local sender receives a local negative acknowledgment or it fails to receive a local acknowledgment, packet A is retransmitted. We define local retransmission delay d_l as the time duration between two consecutive transmissions of the same packet. Let $\tilde{d}_q$ be the waiting time in the new arrival queue and r_t be the number of retransmissions. The total delay jitter d_q equals $\tilde{d}_q + r_t(d_l + 1)$ slots.

For one flow with $W \geq R$ and sufficient buffer, the probability of delay jitter exceeding threshold D can be calculated as follows:

$$\Pr\{d_q > D|W\} = \Pr\{\tilde{d}_q > D|W\}$$

$$+ \sum_{i=W-R+1}^{D-1} \Pr\left\{r_t > \frac{D-i-1}{d_l+1} | \tilde{d}_q = i; \ W\right\}\Pr\{\tilde{d}_q = i|W\}. \tag{6.15}$$

In general, r_t and $\tilde{d}_q$ are independent random variables. r_t can be considered as a geometrically distributed random variable. The packet at the front of the new arrival queue will be transmitted only if the retransmission queue length is less than d_l; therefore, the $\Pr\{\tilde{d}_q > D|W\}$ is equivalent to the probability that fewer than $W - d_l - 1$ packets are successfully transmitted in $R + D$ slots. Equation (6.15) can be rewritten as

$$\Pr\{d_q > D|W\} = \sum_{i=W-R+1}^{D-1} \left(1 - \sum_{k=0}^{\lfloor (D-i-1)/(d_l+1)\rfloor} p_e^k(1-p_e)\right)$$

$$\times \sum_{j=0}^{W-d_l-2} (T_{R+i-1}(j) - T_{R+i}(j)) + \sum_{x=0}^{W-d_l-2} T_{R+D}(x). \tag{6.16}$$

Similarly, we can obtain the delay jitter distribution for the limited buffer case:

$$\Pr\{d_q > D|W; B\} = \Pr\{\tilde{d}_q > D|W; B\}$$

$$+ \sum_{i=W-R+1}^{D-1} \Pr\left\{r_t > \frac{D-i-1}{d_l+1}|W; B\right\}\Pr\{\tilde{d}_q = i|W; B\},$$

(6.17)

where

$$\Pr\{\tilde{d}_q > D|W; B\} = \sum_{x=W-R+1}^{B} \sum_{i=0}^{x-d_l-2} T_{D+1}(i)T_{R-1}(W-x).$$

For the multiple-flow case, the delay jitter distribution is given by

$$\Pr\{d_q > D|\vec{W}; B\} = \Pr\{\tilde{d}_q > D|\vec{W}; B\}$$

$$+ \sum_{i=S-R_N+1}^{D-1} \Pr\left\{r_t > \frac{D-i-1}{d_l+1}|\vec{W}; B\right\}\Pr\{\tilde{d}_q = i|\vec{W}; B\},$$

where

$$\Pr\{\tilde{d}_q > D|\vec{W}; B\} = \sum_{x=S-R_N+1}^{S} \Pr\{q = x|\vec{W}\} \sum_{j=0}^{x-d_l-2} T_{D+1}(j)$$

and $S \geq R_N$.

6.4.5   Delay control for window-controlled protocol

In this subsection, we demonstrate how to apply the analytical results to statistically bound end-to-end delay for window-controlled flows.

The delay distributions derived are conditional on W. The delay outage probability is a non-decreasing function of W, as shown in Subsections. 6.4.1–6.4.4. To bound the delay outage probability, W should be bounded.

On the other hand, the wireless link is underutilized when the arrival rate to the BS is less than the average link throughput, so the window size should be large enough for efficient link utilization. Therefore, a suitable window size is needed to yield a good tradeoff between delay performance and link utilization, and it is necessary to determine the maximum window size, $W_{\max}$, which can bound the delay outage probability.

First, we consider the single flow case. The pseudo code to calculate $W_{\max}$ is given in Algorithm 2. When local retransmission delay is considered, the *DelayOutageProbability* in Line 2 of the algorithm should be set according to (6.16).

If $W_{\max} \geq R$, the wireless link can always be fully utilized no matter what the actual wireless packet error rate is; if $R > W_{\max} \geq (1 - p_e)R$, the wireless link can still be efficiently utilized if the actual packet error rate is no less than the estimated p_e; if $W_{\max} < (1 - p_e)R$, the wireless link may be underutilized. The last case may

Algorithm 2 Algorithm to determine W_{max}, single-flow case [118]. Reproduced by permission of ©2007 IEEE.

1: **for** $(W = 0; W \leq R + D; W + +)$ **do**
2:     $DelayOutageProbability + = T_{R+D}(W)$;
3:     **if** $(DelayOutageProbability > Threshold)$ **then**
4:         $W_{max} = W - 1$;
5:         break;
6:     **end if**
7: **end for**

---

occur when the transmission errors are very bursty and the packet error rate is high. In that case, the BS either uses a stronger FEC code and assigns more wireless channels for the connection, or renegotiates with the application to loosen the delay bound or delay outage bound. Although these are very interesting topics, they are not the focus of this book.

When $W_{max} \geq (1 - p_e)R$, the BS sets the buffer size $B = W_{max}$,[3] and informs the sender to set the upper bound of the window size equal to W_{max}.

The algorithm to determine $W_{max\,i}$ for the multiple-flow case is shown in Algorithm 3. Since the window size is always an integer, W_i should be rounded to the nearest integer, as shown in Line 3. The BS then informs the ith flow's sender to set the upper bound of the window size equal to $W_{max\,i}$, and sets the BS buffer size to $B = \sum_{i=1}^{N} W_{max\,i}$.

Algorithm 3 Algorithm to determine W_{max}, N-flow case [118]. Reproduced by permission of ©2007 IEEE.

1: **for** $(W_1 = 0; W_1 < R_1 + D; W_1 + +)$ **do**
2:     **for** $(i = 2; i \leq N; i + +)$ **do**
3:         $W_i = Round[P_i(W_1/P_1 + E[T_{R_i-1}] - E[T_{R_j-1}])]$;
4:         calculate $DelayOutageProbability$ according to (6.12)
5:         **if** $(DelayOutageProbability \leq Threshold)$ **then**
6:             **for** $(i = 1; i \leq N; i + +)$ **do**
7:                 $W_{max\,i} = W_i$;
8:             **end for**
9:         **else**
10:             break;
11:         **end if**
12:     **end for**
13: **end for**

---

[3]This is achievable, e.g., the Enhanced GPRS system can set the buffer size up to 48KB [135].

6.4.6   Further discussion

In this section, we further discuss the assumptions made in the previous sections.

Delay variation in wireline network

Because of queueing, delay in the wireline domain may also be time varying. For the single-flow case, let R and d_q be respectively the minimal *rtt* and the BS queueing delay used in the calculation, and R' and d_q' be the actual values experienced by a token. When $R' \neq R$, there are two cases. Case (a): When the token returns to the BS, the BS queue is non-empty. In this case, the actual round trip delay $(R' + d_q')$ of the token is equal to the time to transmit W packets successfully over the wireless link, i.e., $\Pr\{d_q' + R' > R + D|W\} = \sum_{x=0}^{W-1} T_{R+D}(x) = \Pr\{d_q > D|W\} = \Pr\{d_q + R > R + D|W\}$. Thus, the actual end-to-end delay jitter $R' + d_q' - R$ has the same distribution as that of d_q, so our analytical results are still valid.[4] Case (b): When the token returns to the BS, the BS queue is empty. In this case, the wireless link is underutilized, and the delay variation in the wireline domain cannot be absorbed by the BS queue. Since the delay in highly-multiplexed wireline networks is not under the control of the end-systems, case (b) is beyond the scope of this book.

Recent Internet measurement results show that the delay jitter in the wireline backbone networks is well controlled [136–138]. Although theoretically the IP-based Internet cannot guarantee delay jitter, in reality, the major ISPs in North America have claimed to guarantee delay jitter in the granularity of milliseconds in their backbone networks, according to their Service Level Agreements [139–141]. Given that a majority of the Internet traffic is well behaved (under TCP or TCP-friendly congestion control), 'bandwidth over-provisioning', an engineering solution widely deployed in today's Internet, makes it possible to successfully support real-time traffic with guaranteed delay. For instance, the measurement results of the AT&T backbone network made in February 2002 show that 'most of the time, the delay variation is undetectable (less than 2 ms per test)' [136]. Other measurement results also show that the networks keep on evolving to provide better quality of service [137, 138]. Due to the comparatively low-cost optical bandwidth, delay guarantee over the Internet through 'bandwidth over-provisioning' has been partly achieved and future growth of the Internet will make it even more viable. As we have witnessed, there are more and more multimedia applications (such as VoIP and IPTV), which have very stringent delay requirements, and it is anticipated that they will occupy a large share of voice communications in the future.

However, due to the limitation in wireless bandwidth, bandwidth over-provisioning is too expensive as a solution for QoS support in wireless cellular systems. Instead, centralized resource management schemes have been deployed to

---

[4] It is worth pointing out that d_q' and R' are *dependent* random variables in this case, so the probability distribution of the end-to-end delay is *not* the convolution of the distributions of d_q' and R'.

Table 6.3   Link parameters.

Link		Propagation delay (slot)	Capacity (packets per slot)
sr_1	duplex	p_0	C_0
r_1r_2	duplex	p_1	C_1
r_2d	duplex	p_2	C_2

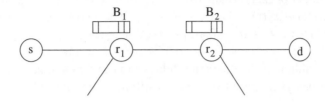

Figure 6.7 Example of window-controlled connection.

maintain the QoS of existing and handoff calls. Therefore, the bottleneck of a cross-domain connection is the wireless access link.

On the other hand, for window-controlled flows, per-node (or per-hop) delay guarantee is not the most efficient way to guarantee end-to-end delay, as demonstrated in the following example.

As shown in Figure 6.7, let a window-controlled connection be established between nodes s and d. All links between s and d are error-free. Links sr_1 and r_2d are dedicated to this flow and cross-traffic will share link r_1r_2 with the target flow. Buffer sizes at r_1 and r_2 are B_1 and B_2, respectively. The capacity C_2 of link r_2d is much less than C_0 and C_1. Drop-Tail queue management is used. The maximum tolerable delay from s to d is $D + p_0 + p_1 + p_2 + 1/C_2$, where D is the maximum tolerable delay jitter.

Proposition 5 *Given a per-node delay guarantee scheme, the queueing delays in r_1 and r_2, d_1 and d_2 should satisfy the following inequalities: $d_1 \leq d_{1\,max}$, $d_2 \leq d_{2\,max}$ and $d_{1\,max} + d_{2\,max} \leq D$.*

However, since the target flow is controlled by the sender window W, d_1 and d_2 are correlated random variables. Let the buffer sizes $B_1 = DC_1$ and $B_2 = DC_2$ (i.e., $d_{1\,max} = d_{2\,max} = D$). We can still guarantee $d_1 + d_2 \leq D$ if the window size $W \leq W_0 + DC_2 + 1$, where $W_0 = 2C_2(p_0 + p_1 + p_2)$.

Proof. At time instant t, a packet from the target flow arrives at r_1. Let $q_1(t)$ and $q_2(t)$ represent the queue lengths in r_1 and r_2 at time t, respectively. Let $x_1(t)$ and $x_2(t)$ denote the number of packets belonging to the target flow in r_1 and in-flight in link r_1r_2 at time t, respectively. Since the end-to-end delay is a non-decreasing function of W, we only need to prove $d_1 + d_2 \leq D$ when $W = W_0 + DC_2 + 1$.

At any time instant τ, $d_1(\tau) = q_1(\tau)/C_1$, where $q_1(\tau) \leq B_1$. Since $d_1 \leq B_1/C_1 = D$, all packets buffered in r_1 arrive within D slots. For the window-controlled target flow, packets arrive at r_1 with a rate less than C_2 packets per slot; therefore, the number of packets from the target flow buffered in r_1, $x_1(\tau)$ is no larger than DC_2.

Since $x_1(\tau) + q_2(\tau) + 2(p_0 + p_1 + p_2)C_2 \geq W > W_0 + DC_2$ and $x_1(\tau) \leq DC_2$, $q_2(\tau) > 0$. In other words, link r_2d is always busy.

At time $t + q_1(t)/C_1 + p_1$ the target packet arrives at r_2; the queue length at r_2 (including the target packet) equals $q_2(t) + x_1(t) + x_2(t) - (d_1 + p_1)C_2$. Therefore, $d_2 = [q_2(t) + x_1(t) + x_2(t) - 1]/C_2 - d_1 - p_1$. The end-to-end queueing delay of the target packet is $d_1 + d_2 = [q_2(t) + x_1(t) + x_2(t) - 1]/C_2 - p_1$. Since $W = q_2(t) + x_1(t) + x_2(t) + (2p_0 + p_1 + 2p_2)C_2$, $d_1 + d_2 = (W - W_0 - 1)/C_2 = D$.

In summary, although the queueing delays in r_1 and r_2 can approach D, the sum of the queueing delays in these two nodes is still no larger than D, if the window size $W \leq W_0 + DC_2 + 1$.

Delayed and lost acknowledgments

To reduce the overhead of *acks*, the receiver may use the delayed *ack* algorithm [142], where an *ack* should be generated for every two packets. The perturbation resulting from delayed *acks* on the delay distribution is equivalent to the case that the R' experienced by some packets is slightly larger than R, as discussed earlier.

For the lost *ack* case, since *acks* can usually acknowledge a number of data packets, with the redundancy in *acks*, the sender can recover the information of an occasional lost *ack* from subsequent *acks*. Thus, the perturbation incurred by lost *acks* is equivalent to that caused by the delayed *ack*.

Unsaturated sender

If the sender does not always have data to send, the number of in-flight packets (tokens) may be less than the window size, so that the actual end-to-end delay outage rate may be less than the analytical result, and the *rwnd*-based delay control scheme can still bound the delay outage rate.

Packet losses due to failed retransmissions by ARQ

If the wireless link is in a bad condition for a long time, some packets may be discarded due to failed retransmissions by the link-level low-persistent ARQ. The packet losses can be viewed as a 'congestion' indicator and they will trigger the sender to exponentially decrease its window size. Also, when some packets are discarded, the queueing delay of those packets following the discarded ones will be less. Thus, if packet losses due to failed retransmissions are considered, the packet loss rate of the target flow is increased, but the delay outage rate will be less than the analytical result, and the *rwnd*-based delay control scheme can still be used to bound the delay outage rate.

Wireless channel model

We use the M-state Markov model and the *persistent* and *CSD* packet transmission schemes to calculate the wireless link throughput distribution. With other wireless channel models and packet transmission schemes, the delay analysis and the control scheme are still applicable, so long as the link throughput distribution can be obtained.

6.5 Parameter Selection for AIMD

To efficiently support delay-sensitive applications in wireless Internet, the design objectives of transport layer protocols are (a) to be TCP-friendly in wired links, (b) to maximize the wireless link utilization and flow throughput, and (c) to provide satisfactory end-to-end QoS.

6.5.1 TCP-friendliness

As shown in Chapter 4, AIMD parameters satisfying the following condition can guarantee TCP-friendliness, no matter what the bottleneck link capacity is and how many TCP and AIMD flows coexist in the link:

$$\alpha = 3(1 - \beta)/(1 + \beta), \tag{6.18}$$

where $0 < \alpha < 3$ and $0 < \beta < 1$. In addition, the DTAIMD algorithm proposed in Chapter 4 can mitigate the practical implications to enhance TCP-friendliness. Different applications can choose one of the parameters, and the other parameter is determined by (6.18). However, the DTAIMD algorithm still tends to bias against long rtt flows, as TCP does, unless the windows of the coexisting flows satisfy (6.9).

6.5.2 *rwnd*, single AIMD flow

With the AIMD congestion control mechanism, AIMD flows probe for available bandwidth and overshoot the bottleneck link capacity frequently, which produces transient congestion and packet losses. For a highly multiplexed bottleneck, dynamic probing with the AIMD mechanism is required since the end-systems do not have the knowledge of the global traffic. However, for a cross-domain connection, the bottleneck is most likely the lightly multiplexed wireless link, and, in general, dedicated wireless links are allocated to multimedia flows. The dynamics of available bandwidth in wireless links may not be due to the competition of multiplexed flows, but due to the time-varying wireless channel condition. Overshooting a dedicated wireless link frequently is not an efficient way to utilize it.

As shown in Subsection 6.4.1, the sufficient condition to maximize the wireless link utilization and flow throughput is $W \geq R$, where $W = \min(rwnd, cwnd)$. Obviously, $rwnd \geq W$. Less obvious is the fact that we can choose the appropriate $rwnd$ and B such that W converges to $rwnd$, i.e., $W = rwnd \geq R$.

When $B < W$, if the number of successful transmissions in $R - 1$ slots is less than $W - B$, buffer overflow at the BS may occur. Packet losses due to poor channel conditions can trigger the AIMD sender to exponentially reduce W. Consequently, when the channel condition becomes better later, there may not be enough packets for transmission due to the small window size. Thus, the wireless link buffer size, B, can be conservatively set to *rwnd* to avoid buffer overflow due to wireless channel dynamics; the exponential backoff will not be triggered unnecessarily.

Ideally, if there is no delay jitter and packet loss in the wireline domain, by setting $B = rwnd = R$ the AIMD sender window size converges to *rwnd* in steady state, i.e., $W = rwnd = R$. Thus, the wireless link is fully utilized and the flow throughput is maximized, and the minimal queue length at the BS is 1.

In reality, there may be delay jitter and packet losses in the wireline domain. With $B = rwnd$, end-to-end packet loss rate equals the packet loss rate in the wireline domain, which can be estimated and bounded based on prior measurements (e.g., using the data in [137]). On the other hand, setting *rwnd* to R cannot absorb delay jitter in the wireline domain, so that the wireless channel may be idle sometimes. For instance, when $rwnd = R$, if one packet is delayed one more slot in the wireline domain, the BS queue may be empty for one slot. Such underutilization of the wireless link can be avoided if $rwnd > R$. Also, when a single packet loss occurs in the wireline network, the AIMD sender will exponentially reduce its sender window. If β times *rwnd* is less than R, the queue at the BS will be empty for some slots, and the wireless link is underutilized. Therefore, to efficiently utilize the wireless link, it is better to set a larger *rwnd*.

However, a larger *rwnd* leads to a larger queue length and higher delay outage probability, since the delay outage probability given by (6.3) is a non-decreasing function of the window size. The maximal window size, W_{max}, satisfying the maximum tolerable delay outage probability is the optimal value of *rwnd*, which can efficiently utilize the wireless link while statistically guaranteeing the delay outage probability. The algorithm to derive W_{max} has been given in Algorithm 2.

6.5.3   *rwnd*s, multiple AIMD flows

For multiple AIMD flows sharing the wireless link, to assign a ratio of wireless bandwidth to AIMD flows with the same or different *rtt*s, their window size should satisfy (6.9). The BS buffer size can be set to S to avoid buffer overflow, so the end-to-end packet loss rate equals the delay outage rate plus a constant (the estimated packet loss rate in the wireline domain). The delay outage probability is given by (6.12). Under the *necessary* and *sufficient* constraint of (6.9), the optimal *rwnd*s can be set to the maximum integers satisfying the delay outage probability, according to Algorithm 3.

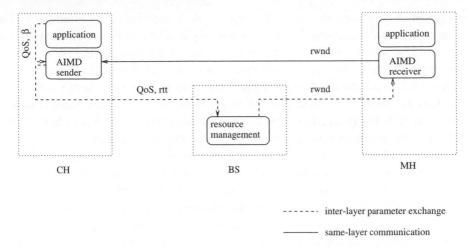

Figure 6.8 Parameter exchange [82]. Reproduced by permission of ©2006 IEEE.

6.5.4 Parameter selection procedure

To achieve the performance attainable by the AIMD congestion control protocol, the parameter values should be appropriately selected and/or calculated, and distributed. The procedure to perform the parameter value selection/calculation is as follows:

1. The application identifies its required throughput, maximum tolerable delay jitter D, delay outage probability, and packet loss rate; it also selects a desired value of β according to its maximum tolerable throughput variation.

2. The application passes these parameter values to the transport layer protocol.

3. In the transport layer, for a given value of β, the AIMD sender calculates α according to the TCP-friendly condition (6.18), and measures and estimates the minimal rtt of the flow.

4. The AIMD sender sends the minimal rtt and the QoS parameters to the BS.

5. According to the wireless link characteristics and the physical layer protocol, the BS resource allocation module (in the link layer) calculates the packet-level wireless channel profile, determines the link layer transmission scheme and allocates wireless channels to the connection such that the average link throughput is no less than the desired flow throughput.

6. The BS calculates W_{max}, the optimal $rwnd$, according to the QoS requirements and the wireless link throughput distribution, and sets $B = W_{max}$.

7. The BS informs the AIMD receiver to set the allocated buffer size to W_{max}. Then, the AIMD receiver informs the sender (by *ack*) that the *rwnd* is W_{max}.

The exchange of parameter values in steps 2, 4, 7 are shown in Figure 6.8. The computations involved in steps 3 and 5 are very light. The computations in step 6 can be done offline and the results can be stored in a look-up table, so the BS can check the table to determine the appropriate values of *rwnd* and *B* quickly. Therefore, the parameter value selection/calculation procedure can be done efficiently during the connection establishment phase. Steps 5 to 7 will be repeated when handoff occurs, or when the wireless channel profile changes significantly (as shown in Section 6.6, optimal *rwnd* is insensitive to small changes of the channel profile).

To efficiently support multimedia applications with heterogeneous QoS requirements in hybrid wireless and wireline networks, inter-layer interactions are necessary. Nevertheless, such interactions should be minimized for scalable system design purposes. As shown in the above procedure, the parameter value selection/calculation only requires the (infrequent) exchange of QoS and protocol parameters among the application, the transport layer protocol and the link layer protocol. Therefore, the proposed approach preserves the end-to-end semantics of the transport layer protocol and the layered structure of the Internet, and is applicable to supporting various multimedia applications with a wide variety of QoS requirements over wireless links with different channel models and physical/link layer protocols.

6.6   Performance Evaluation

To demonstrate the feasibility of the proposed AIMD protocol, examine link utilization and evaluate the protocol performance, we have performed extensive simulations using the NS-2 simulator.

6.6.1   Single AIMD flow

The simulation topology is the same as that in Figure 6.1. For the target AIMD-controlled multimedia flow, the sender is at the CH and the receiver is at the MH. The MH and the BS are connected to routers r_1 and r_2, respectively. Cross traffic connections share the r_1r_2 backbone link. The following parameter values are used in the simulation unless otherwise explicitly stated.

Wired links between the CH, r_1, r_2, and the BS are duplex links with 100 Mbps. The propagation delay from the CH to the MH is 37.5 ms. Both r_1 and r_2 are RED capable, with the following parameter values: minimum threshold 50 packets, maximum threshold 250 packets, and maximum packet dropping probability 0.1 (as the average queue size varies between the two thresholds). The downlink and uplink between the BS and the MH have the capacity of 200 Kbps and 100 Kbps, respectively. The downlink channel condition is dynamically changed according to an M-state Markov model. The target multimedia flow has the packet size of 125 bytes. Thus, the duration of a time slot is 5 ms. The deterministic end-to-end delay of the target flow is 42.5 ms. The AIMD protocol uses the parameter values, $\alpha = 0.2$ and $\beta = 0.875$, according to the TCP-friendly condition. Cross traffic sharing r_1r_2

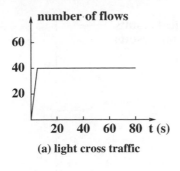
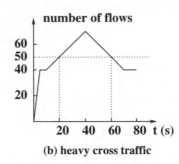

(a) light cross traffic   (b) heavy cross traffic

Figure 6.9   Number of cross TCP connections in r_1r_2 [118]. Reproduced by permission of ©2007 IEEE.

are TCP-SACK flows with a packet size of 1250 bytes. To eliminate phase effect, the *rtt*s of the cross TCP-SACK flows are randomly chosen between 80 ms and 90 ms.

Each simulation lasts 80 s, and different initial randomization seeds are used to reduce simulation dynamics. To eliminate the system warming-up effects, simulation results for the first 5 s are not counted.

Delay outage rate

To examine the effectiveness of the analysis in various scenarios, simulations with light cross traffic and heavy cross traffic are performed separately. The number of cross TCP-SACK connections is adjusted according to Figure 6.9. In the light cross traffic case, the bottleneck for the target AIMD flow is always the wireless link; in the heavy cross traffic case, the bottleneck is the backbone link between 20 s and 60 s when there are more than 50 TCP connections in r_1r_2.

The delay outage rate is the ratio of the number of received packets with the end-to-end delay exceeding 92.5 ms (equivalent to the delay jitter exceeding 50 ms) over the number of packets sent. The tolerable delay outage rate is 1%. The BS buffer size is set to *rwnd* to avoid buffer overflow.

With the *persistent* transmission scheme, Figures 6.10(a) and (b) compare the analytical and simulation results of the maximum *rwnd*s (W_{max}) in which the delay outage rate is below 1%, for the light cross traffic and the heavy cross traffic cases, respectively. The unit of W_{max} is packets per *rtt*. Since the *persistent* transmission scheme is suitable for channels with a small λ value, the wireless channel is modeled using a two-state Markov chain with λ equal to zero, and the average packet error rate (p_e) varies from 0 to 0.1. The local retransmission delay is 5 ms.

The analytical and simulation results with the CSD transmission scheme are shown in Figures 6.11(a) and (b). The wireless channel is modeled using a three-state Markov chain with λ equal to 0.35. According to the CSD transmission scheme

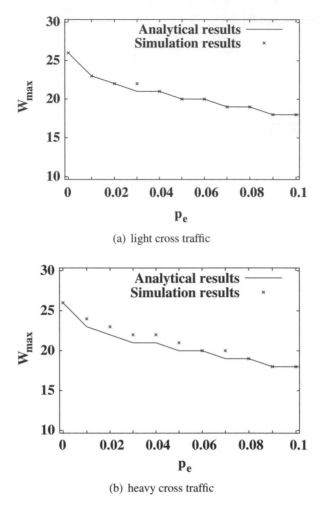

(a) light cross traffic

(b) heavy cross traffic

Figure 6.10 W_{max}, *persistent* transmission scheme [82]. Reproduced by permission of ©2006 IEEE.

in [108], the BS suspends transmission for one slot if the previous transmission fails, and there is no local retransmission delay.

The analytical value of W_{max} is calculated according to Algorithm 2. Figures 6.10 and 6.11 show that the analytical results match well with the simulation ones when the cross traffic is light (when the bottleneck is always the wireless link). The analytical ones are slightly more conservative when the cross traffic is heavy. With heavy cross traffic, the bottleneck link is the backbone link between 20 s and 60 s, so that *cwnd* and W are smaller than *rwnd* during this period, and the delay outage rate will be smaller. On the other hand, when r_1r_2 becomes more congested, queueing delay in the wireline domain becomes more significant. These two effects (smaller

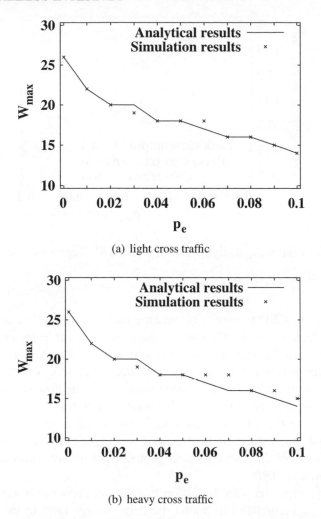

(a) light cross traffic

(b) heavy cross traffic

Figure 6.11 W_{max}, CSD transmission scheme [82]. Reproduced by permission of ©2006 IEEE.

sender window and larger delay in the wireline domain) offset each other, so that the heavy and light cross traffic cases have similar delay outage rates.

Another observation is that W_{max} changes slowly w.r.t. p_e. For instance, when p_e varies from 5% to 8%, W_{max} only changes from 20 to 19 for the *persistent* transmission cases, and from 18 to 16 for the CSD transmission cases. Therefore, the *rwnd*-based control scheme can tolerate a certain degree of error in estimating the wireless channel profile.

Comparing the results in Figures 6.10 and 6.11, the W_{max} with the CSD transmission scheme is smaller than that with the *persistent* transmission scheme.

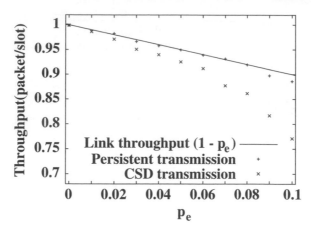

Figure 6.12 Flow throughput, light cross traffic [118]. Reproduced by permission of ©2007 IEEE.

This is because the CSD transmission scheme makes a tradeoff in achieving multi-user gain and higher power efficiency at the cost of possible lower throughput of a particular flow. In the simulations, only a single wireless link is simulated, and the interferences between wireless links are not addressed. Therefore, the multi-user gain is not reflected in the simulation results, and the flow throughput with the CSD transmission scheme is less than that with the *persistent* transmission scheme, as shown in Figure 6.12. Flow throughput is measured in terms of the number of packets received successfully (in the transport layer) per time slot. Since our focus is on the performance of the transport layer protocol, we do not consider the link layer optimization problem here.

In summary, no matter what link layer transmission scheme is used, it is feasible to calculate W_{max} beforehand to bound the delay outage rate. In the following, we present simulation results for *persistent* transmission only.

Link utilization and TCP-friendliness

Ideally, for flow and congestion control in hybrid wireless and wireline networks, the allocated wireless resources should be fully utilized if the bottleneck is the wireless link, and the cross-domain AIMD flow should occupy its fair share of bandwidth only when the bottleneck is the highly multiplexed backbone link, i.e., being TCP-friendly.

For the wireless link utilization, W_{max} should be larger than $R(1 - p_e)$ in order to efficiently utilize the wireless link. Our analytical and simulation results show that W_{max} is larger than $R(1 - p_e)$ in most cases except for the CSD transmission scheme with $p_e = 0.1$. Therefore, W_{max} does not limit the flow throughput to underutilize the wireless link in most cases. Figure 6.12 also demonstrates the efficiency of the *rwnd*-controlled scheme: with *persistent* transmission and light cross traffic, by setting

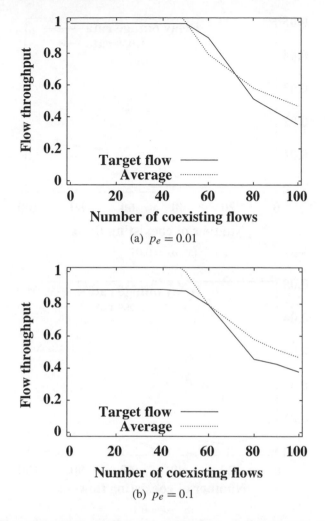

(a) $p_e = 0.01$

(b) $p_e = 0.1$

Figure 6.13 Flow throughput: (a) *rwnd* = 23; (b) *rwnd* = 18 [82]. Reproduced by permission of ©2006 IEEE.

rwnd to the calculated W_{max}, the flow throughput (packet per slot) is close to the average link throughput.

To further examine the wireless link utilization and TCP-friendliness, the number of coexisting TCP-SACK flows (*elephants*) in the backbone link is changed from 0 to 100. The packet size of the cross TCP-SACK flows is 1250 bytes. Figures 6.13(a) and (b) plot the throughputs of the target AIMD flow and the average throughputs of all coexisting TCP and AIMD flows, with p_e equal to 0.01 and 0.1, and *rwnd* equal to 23 and 18, respectively. When the number of coexisting flows in the backbone link is less than 50, the average throughput in the backbone link is larger than one packet

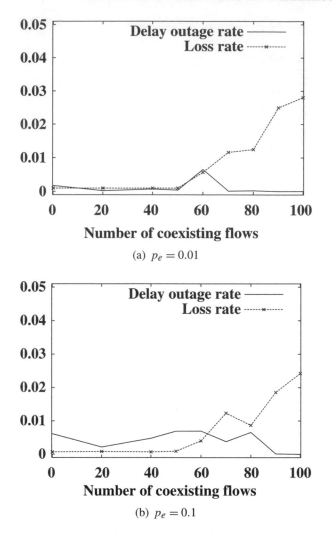

(a) $p_e = 0.01$

(b) $p_e = 0.1$

Figure 6.14 Delay outage rate and packet loss rate [82]. Reproduced by permission of ©2006 IEEE.

per slot, and the bottleneck for the target AIMD flow is the wireless link. Therefore, the throughput of the target flow should be $(1 - p_e)$ to fully utilize the wireless link. When the number of coexisting flows is larger than 50, the backbone link becomes the bottleneck, and the throughput of the target AIMD flow should be close to the average throughput in the backbone link in order to be TCP-friendly. It can be seen that the target AIMD flow demonstrates the desired performance. Simulation results with other values of p_e show a similar tendency.

Meanwhile, as shown in Figure 6.14, the delay outage rate of the target flow meets the requirement of $\leq 1\%$. The packet loss rates are negligible when the number of coexisting flows is less than 50, and packet loss rates increase to 2.8% when the number of coexisting flows is increased to 100, which indicates that the packet losses are mainly due to network congestion in the wireline domain. Therefore, with appropriate buffer size and window size, the packet loss over the wireless domain and the delay outage rate can be bounded.

6.6.2 Multiple AIMD flows

The simulation topology for multiple AIMD flows sharing a wireless link is similar to that shown in Figure 6.1, except that there are N AIMD senders at N CHs connected to r_1.

Let two AIMD flows have the same share of wireless resources. According to (6.9), their sender windows should satisfy $W_1 = W_2 + (E[T_{R_1-1}] - E[T_{R_2-1}])/2$. Also, to bound the delay outage rate, $\Pr\{d_q > D|\vec{W}\}$ should be less than 0.01. The rwnds are set as the maximum integers satisfying the above two constraints. The link buffer size is set equal to the sum of rwnds to avoid buffer overflow at the BS. The cross TCP-SACK traffic is set according to Figure 6.9(a). Figure 6.15(a) shows the throughputs and delay outage rates of two flows with the same minimal rtts of 17 slots, w.r.t. p_e. In Figure 6.15(b), two flows have different rtts: $R_1 = 17$ and $R_2 = 25$. The simulation results demonstrate that the coexisting AIMD flows can fairly share the wireless link, and satisfy the delay outage bounds, no matter whether they have the same value of rtt or not.

The same conclusion can be drawn when the number of flows sharing the wireless link is increased. Figures 6.16(a) and (b) show the throughputs and the delay outage rates for four coexisting AIMD flows. These four flows are designed to occupy the same share of wireless resources. In Figure 6.16(a), all four flows have the same minimal rtts of 17 slots; in Figure 6.16(b), $R_1 = R_3 = 17$ and $R_2 = R_4 = 25$.

Next, let AIMD flows have different shares of the wireless link. Figure 6.17 shows the performance of two AIMD flows with minimal rtts of 17 slots and 25 slots. Flow 1 is assigned 1/3 share of the wireless link, and flow 2 is assigned 2/3 share. The simulation results demonstrate that the coexisting AIMD flows can achieve their fair shares of the wireless link and satisfy the delay outage bounds, even when they have different rtts and have been assigned different shares of the wireless link.

Since rwnd is an integer, we may not be able to get a group of rwnds to satisfy (6.9) exactly. Therefore, sometimes the results slightly deviate from our designed target due to quantization errors, e.g., in some cases the throughputs of coexisting AIMD flows have a small difference, and in others the delay outage rates exceed the desired 1% slightly. Nevertheless, such deviations can be anticipated and controlled.

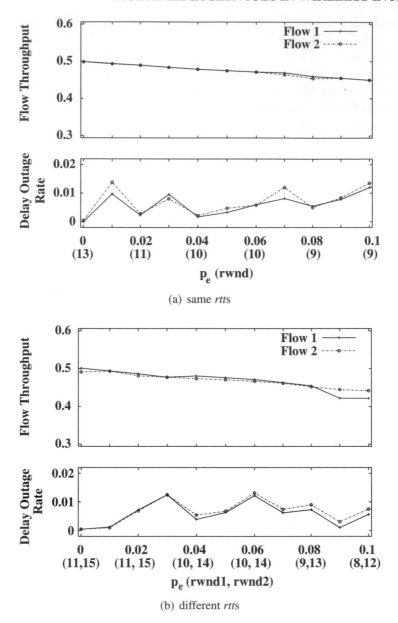

(a) same *rtt*s

(b) different *rtt*s

Figure 6.15 Two AIMD flows sharing the wireless link [118]. Reproduced by permission of ©2007 IEEE.

6.6.3 AIMD vs. TCP

We further evaluate the performance of AIMD flows with *rwnd* equal to W_{max} and TCP flows with *rwnd* equal to 50 packets. Figure 6.18 compares the throughputs and

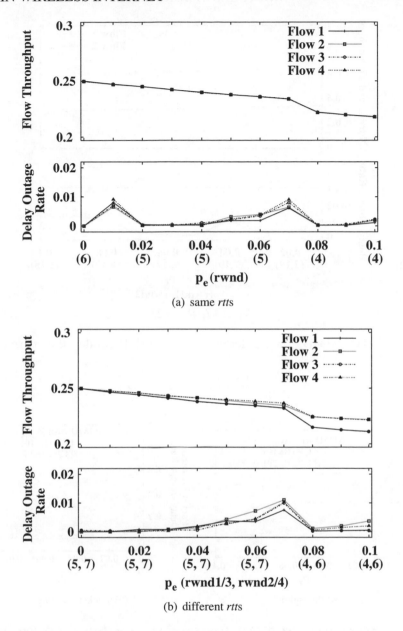

Figure 6.16 Four AIMD flows sharing the wireless link [118]. Reproduced by permission of ©2007 IEEE.

packet loss rates (due to both buffer overflow and excessive delay) of the TCP and AIMD flows w.r.t. p_e. For the TCP flows, we set the BS buffer size to 10 and 20, respectively. Figure 6.18(a) shows that if the buffer size is larger, the throughputs

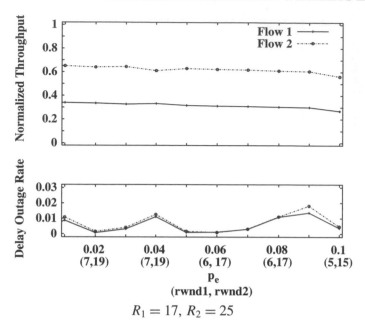

$$R_1 = 17, R_2 = 25$$

Figure 6.17 Two AIMD flows, different shares [118]. Reproduced by permission of ©2007 IEEE.

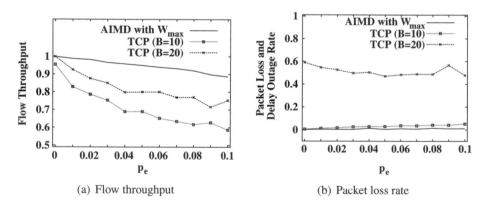

(a) Flow throughput   (b) Packet loss rate

Figure 6.18 AIMD vs. TCP [82]. Reproduced by permission of ©2006 IEEE.

of TCP flows can be higher; but with a larger buffer, the delay outage probability increases significantly, as shown in Figure 6.18(b). In addition, no matter which buffer size is chosen, the throughputs of TCP flows are less than those of AIMD flows, and the packet loss rates of TCP flows are higher than those of AIMD flows.

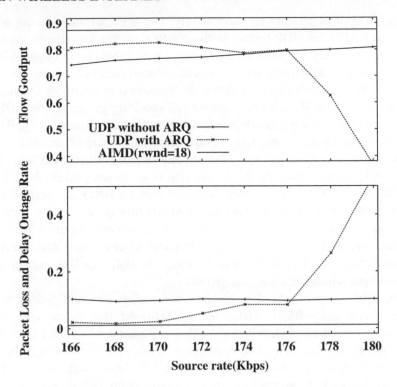

Figure 6.19 AIMD vs. UDP, $p_e = 0.1$ [82]. Reproduced by permission of ©2006 IEEE.

This figure further demonstrates the advantages of the proposed AIMD protocol and its parameter selection scheme.

6.6.4 AIMD vs. UDP

The Internet has evolved from a small, research-oriented, cooperative system to an enormous, commercial, competitive information transport infrastructure. From the users' point of view, they would like to discard any congestion control in their systems if such control has negative effects on their perceived QoS. How to punish greedy and malicious users is beyond the scope of this book. However, by appropriately choosing the protocol parameters, the responsive AIMD protocol can outperform the unresponsive UDP protocol when supporting multimedia applications over wireless networks. This can be an incentive for the end-systems to deploy AIMD congestion control.

Since the UDP protocol has no closed-loop control mechanism, the sender just keeps on sending at the source rate. Assume that the UDP sender still has the knowledge of the wireless link, e.g., link bandwidth, p_e, and the UDP sender can determine an optimal sending rate accordingly. The BS can choose to use or not use the link level ARQ for UDP traffic. Without ARQ, packet loss due to transmission error is approximately p_e, which may be too severe when p_e is large. With ARQ, packet losses due to transmission errors can be reduced significantly. However, unlike the AIMD sender, which can slow down the transmission when the channel is in a poor condition due to its acknowledgment self-clocking property, the UDP sender cannot change the sending rate adaptively. Consequently, the queue at the BS is built up, and the queueing delay and delay outage probability increase quickly.

In the simulation, since the maximum tolerable delay jitter is ten slots, to avoid excessive delay outage rate, the BS buffer size is set to ten packets for UDP flows with CBR sources. The wireless links with and without ARQ are used for the UDP traffic, respectively. As a comparison, let an AIMD flow over the same wireless link with the link-level ARQ, and the BS buffer size is set to *rwnd* for the AIMD flow. We define the goodput of a flow as the number of packets being successfully received by the transport layer receiver with tolerable delay per slot, which is equivalent to the flow throughput minus the delay outage rate.

Figure 6.19 compares the flow goodput and packet loss and delay outage rate for UDP- (with and without ARQ) and AIMD-controlled flows, with $p_e = 0.1$. Simulation results show that, without ARQ, the packet loss rate for the UDP flow is approximately 0.1; this makes reconstruction of multimedia streams at the receiver difficult. With ARQ, the source rate of the UDP flow should be less than the average wireless link throughput to avoid excessive delay outage rate. As shown in Figure 6.19, no matter how the UDP sender adjusts its sending rate, and no matter whether the wireless link deploys the ARQ for UDP traffic or not, the goodputs of the UDP flows are consistently lower than that of the AIMD flow, and the packet loss and delay outage rates of the UDP flows are consistently higher than that of the AIMD flow. This is because the UDP sender does not receive any feedback about the network/link condition, so it cannot adapt the sending rate accordingly and it easily underutilizes or overshoots the available bandwidth. With the self-clocking property, the AIMD sender can slow down when the wireless link experiences bad channel conditions, as it takes longer for the *ack*s to reach the sender.

In summary, from the network service provider's point of view, the unresponsive UDP protocol may endanger network stability and integrity; from the users' point of view, the window-based AIMD protocol can provide better QoS in terms of higher flow goodput and lower packet loss rate.

6.7 Summary

This chapter introduces an analytical framework for quantifying the performance of window-controlled flows in hybrid wireless and wireline networks. The analytical results can be used to provide timely delivery services for delay-sensitive multimedia applications in wireless Internet. We use the TCP-friendly AIMD protocol proposed in Chapter 4 for supporting delay-sensitive multimedia applications over hybrid networks as an example. By appropriately selecting the protocol parameters, wireless bandwidth can be efficiently utilized and flow throughput can be maximized, under the constraint that the delay outage probability is statistically bounded. Simulation results have validated the analysis, demonstrated the feasibility of the approach and shown that the AIMD protocol can outperform the unresponsive UDP protocol when they are used to support multimedia applications in hybrid networks.

6.8 Problems

1. A TCP sender transmits a large file over a link with 10 Mbps. No other traffic shares the link with it. Assume that there is no transmission error, and the routers' buffer size is much smaller than the file size. Is it possible that there is no packet loss for the whole file transfer? If so, how?

2. Let a window-controlled flow occupy a wireless link. Assume that all packets are of the same size and the wireless link has a buffer of 20 packets and uses the *persistent* transmission scheme. The round-trip time without the queueing delay for the flow is 50 ms. It takes 5 ms to transmit a single packet. The wireless channel is modeled as a discrete time two-state Markov chain with slot duration of 5 ms. In the good channel state, all transmissions are successful, and in the bad state, all transmissions are failed. A corrupted packet will be retransmitted immediately. The state transition probabilities from the good state to the good state and from the bad state to the bad state are 0.95 and 0.05, respectively.

   (a) What is the memory of the Markov chain for the wireless channel?

   (b) What is the average throughput of the wireless link in terms of packets per second?

   (c) What is the minimum window size such that the wireless link is fully utilized?

   (d) What is the maximum window size such that there is no packet loss due to buffer overflow?

   (e) The delay outage rate is counted as the ratio of the number of received packets with the queueing delay exceeding 50 ms. What is the maximum window size such that the delay outage rate is below 1%?

   (f) Use NS-2 simulation, letting a TCP flow with receiver window size set to be the window sizes obtained above, to verify the analysis.

3. A bursty wireless channel is modeled as a discrete-time two-state Markov chain with slot duration 3 ms. The channel state transition probabilities from the good state to the good state and from the bad state to the bad state

are 0.9936 and 0.36, respectively. In the good state, all transmissions are successful, and in the bad state, all transmissions are failed. The wireless link deploys the following transmission scheme: if the transmission is successful, a new packet will be transmitted immediately; if the transmission fails, the transmitter will be idle for 3 ms, and retransmit the corrupt packet. Assume all packets have the same size and the transmission time of a packet over the wireless link is 3 ms.

Let a TCP flow with receiver window size 20 packets occupy the link. The round-trip time without queueing delay for the TCP packets is 51 ms.

   (a) What is the memory of the Markov chain for the wireless channel?

   (b) What is the minimum buffer size of the wireless link to ensure that there are no packet losses due to buffer overflow?

   (c) Set the buffer size according to the above calculation. In steady state, what is the probability that the queueing delay of a packet exceeds 30 ms?

4. Let two TCP flows share a bottleneck link, and the round-trip time without queueing delay for the two flows be 68 ms and 100 ms, respectively. All packets have the same size and the transmission time of a packet over the bottleneck link is 4 ms. Assume that there are no transmission errors. How can we let the two flows equally share the bottleneck bandwidth? Use NS-2 simulation to verify your analysis.

5. Consider the following wireless channel model and repeat question 2. The wireless channel is modeled as a discrete time two-state Markov chain with slot duration of 5 ms. In different channel states, the transmission rates are different and the packet error rates in both states are negligible. In the good state and the bad state, it takes 5 ms and 10 ms to transmit a packet, respectively. The state transition probabilities from the good state to the good state and from the bad state to the bad state are 0.95 and 0.05, respectively.

6. To download web pages using a cellphone, the bottleneck is assumed to be in the downlink from the base station to the mobile host. Discuss whether opening multiple TCP connections to download the web pages can improve the download speed or not.

7

TCP-friendly Rate Control in Wireless Internet

7.1 Introduction

The TFRC protocol [143] is a representative equation-based rate control protocol. TFRC uses a stochastic model to estimate the throughput of a TCP flow under the same circumstances and sets the maximum allowable sending rate accordingly. According to the stochastic model, the TCP throughput (r_a) can be expressed as a function of the round-trip time (rtt), retransmission timeout (T_0), and packet loss event rate p [144]:

$$r_a = \frac{1}{rtt\sqrt{2p/3} + T_0(3\sqrt{3p/8})p(1 + 32p^2)}.$$ (7.1)

One or more packet losses during one rtt are counted as one loss event. p is measured and calculated by the TFRC receiver using an average loss interval method. The receiver first calculates the average loss interval, and sets the loss event rate p as the inverse of the weighted average of loss event interval. This is done by using a filter that weights the n most recent loss event intervals in such a way that the measured loss event rate changes smoothly. The weights w_0 to $w_{(n-1)}$ can be calculated by:

  if $(i < n/2)$ **then**
    $w_i = 1$;
  else
    $w_i = 1 - (i - (n/2 - 1))/(n/2 + 1)$;
  end if

The value n for the number of loss intervals used in calculating the loss event rate determines the TFRC's speed in responding to changes in the level of congestion.

Multimedia Services in Wireless Internet: Modeling and Analysis   Lin Cai, Xuemin Shen and Jon W. Mark
© 2009 John Wiley & Sons, Ltd

In practice, a value of $n = 8$ appears to be acceptable, which balances between resilience to noise and responding quickly to changes in network conditions, and the values of w_0 to w_7 are: 1.0, 1.0, 1.0, 1.0, 0.8, 0.6, 0.4, 0.2. rtt and T_0 are estimated and calculated by the TFRC sender using the exponentially weighted moving average (EWMA) method. The implementation recommendations on how to measure and calculate these parameters are given in the protocol specification [145].

The TFRC receiver periodically sends feedback messages to the TFRC sender at least once per rtt or whenever a new loss event is detected. Initially, the TFRC sender sets its rate to one packet per second and doubles the rate every rtt until loss occurs. Thereafter, the sending rate is set according to (7.1).

Since TFRC is less aggressive in probing for available bandwidth, and more moderate in responding to transient network congestion, TFRC can achieve TCP-friendliness, and its sending rate is much smoother than that of TCP. In addition, the TFRC sender has no obligation to retransmit lost or corrupted packets, which is desirable for many time-sensitive applications.

In error-prone wireless networks, similar to TCP, the TFRC sender treats any packet loss as due to network congestion and sets its sending rate inversely proportional to the square root of the loss event rate, which leads to underutilization of the wireless link. The link-level ARQ scheme can effectively improve the TFRC performance in wireless links and it is anticipated to be deployed in wireless networks [107] with the following advantages: (a) the link-level ARQ scheme can efficiently recover transmission errors; (b) it is comparatively simple to deploy; and (c) there are no competitive retransmissions between the TFRC and the link layer, because the TFRC sender is not obligated to retransmit lost packets.

In the following, we introduce an analytical framework for evaluating the performance of TFRC in hybrid wireless and wireline networks [146]. Considering a wireless network with a link-level truncated ARQ scheme and limited interface buffer size, one DTMC is developed to investigate the wireless bandwidth utilization and packet loss rate of TFRC flows over wireless links, and another DTMC is developed to study the delay outage probability. Analytical solutions are invariably obtained by making appropriate assumptions, which can provide only approximate results. The reasonableness of the assumptions has to be verified by constructing the system and making measurements, or by extensive simulation study using simulators such as NS-2. Nevertheless, analytical results do provide useful guidelines for network parameter selection and optimization, such as the interface buffer size and the maximum number of retransmissions of the truncated ARQ scheme. With the optimal parameters, wireless link utilization can be maximized, and the QoS requirements of multimedia applications can be statistically guaranteed.

7.2 System Model

Consider a scenario in which a CH communicates with an MH over a hybrid wireless/wireline network. As shown in Figure 7.1, a TFRC connection is established

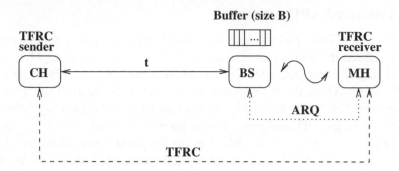

Figure 7.1   TFRC connection in hybrid wireless and wireline networks [146]. Reproduced by permission of ©2006 IEEE.

Table 7.1   Notation for Chapter 7.

Symbol	Description
B	BS buffer size
r	Wireless link capacity (packets per second)
μ	Doppler frequency shifts
l_b	Average length of burst errors
M	Maximum number of transmissions allowed for each packet in the truncated ARQ scheme
p	Loss event rate
p_e	Average packet error rate in transmission
p_l	Packet loss rate
q	Queue length
r_a	TFRC flow throughput (packets per second)
t	Deterministic end-to-end delay
T	Packet transmission time over the wireless link
T_0	Timeout value
t_q	Queueing delay
η	Link utilization
v	Velocity of the mobile host

between a CH and an MH. The MH is connected with a BS through a last-hop wireless link. Assume that all packets of the target TFRC flow have the same packet size and travel through the same route. The BS allocates a dedicated wireless link with a capacity of r packets per second and a buffer of size B for the TFRC flow. End-to-end packet delay contains a deterministic part and a random part; the former represents delay and the latter delay jitter. For easy reference, the notations used in this chapter are listed in Table 7.1.

7.2.1 Truncated ARQ scheme

To improve the TFRC performance over lossy wireless links, a truncated ARQ scheme in which the number of transmission attempts is limited to M is deployed in the link layer. Each packet sent by the TFRC sender (at the CH) is first queued in the BS buffer. Then, the BS forwards the packet at the head of the queue to the MH, where the TFRC receiver resides. In the link layer, if a packet is received by the MH successfully, the MH sends back a local acknowledgment to the BS immediately. Upon receipt of a positive *ack*, the BS removes the packet from its buffer. If the MH fails to receive the packet, the BS retransmits the corrupted packet up to M times. Packets are removed from the buffer after being received by the MH successfully, or after M failed attempts. Assume that there is no packet loss in the wireline domain, and that the packet losses visible to the TFRC receiver are those due to BS buffer overflow and those discarded after M failed attempts.

7.2.2 Wireless channel model

Time is discretized into slots, and the duration of each slot is the transmission time of a packet in the wireless link. Finite-state Markov models are used to characterize wireless channels at the packet level. As discussed in Subsection 2.2.2, a multiple-state Markov model can be mapped to a two-state Markov model. With the two-state model, the channel alternates between a *good* state (in which all packets are transmitted error free) and a *bad* state (in which all transmissions fail). The residence time of each state follows a geometric distribution. Furthermore, the state transition probabilities depend on the normalized Doppler frequency. Let v and μ be the velocity of the MH and the Doppler frequency, respectively. Let T denote the packet transmission time over the wireless link. If μT is small, the channel fading process is highly autocorrelated, and vice versa. Since μ is proportional to v, the slower the MH moves, the burstier are the transmission errors.

7.3 Analytical Model for Rate-controlled Flows

The performance indexes considered are the normalized wireless link utilization, packet loss rate and delay outage rate. The normalized wireless link utilization is defined as the number of packets transmitted successfully per T. Packet loss rate is the ratio of the number of lost packets over the number of packets sent by the TFRC sender. Delay outage rate is the ratio of the number of received packets with delay jitter exceeding a prescribed threshold.

We make the following assumptions to facilitate performance analysis:

1. The bottleneck of the cross-domain connection is the wireless link, and the packet losses due to congestion between the CH and the BS are negligible. This tends to be acceptable, since in general, wired links have larger bandwidth and lower bit error rates than wireless ones.

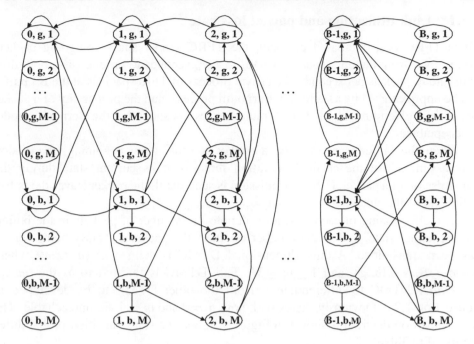

Figure 7.2 State transition diagram of the queue-associated DTMC [146]. Reproduced by permission of ©2006 IEEE.

2. The delay jitter of a packet is mainly due to queueing delay in the BS buffer, since the wireless link is presumably the bottleneck link, and the throughput of the wireless link with ARQ is highly dynamic. Therefore, the delay outage rate equals the probability that a packet's queueing delay at the BS exceeds the maximum tolerable delay jitter.

3. A perfect and instantaneous wireless uplink channel is assumed so that, at the end of each time slot, the BS knows whether or not the transmission is successful (e.g., in a TDMA system, a logical link corresponds to a time slot of each frame, and the time duration between two consecutive slots is long enough to accommodate the link-level acknowledgment [107]). Any packets not successfully received are immediately retransmitted in the next time slot to reduce jitter.

In packet transmissions, the quality of service factors of interest are link utilization, packet loss rate and packet delay. In what follows, we use a DTMC to model the TFRC flows to facilitate analytical evaluation of link utilization, packet loss rate and packet delay. In particular, we introduce the concepts of *queue-associated* DTMC and *packet-associated* DTMC in Subsections 7.3.1 and 7.3.2, respectively.

7.3.1 Link utilization and packet loss rate

Since TFRC is a rate-based protocol, the TFRC sender spaces outgoing packets evenly by setting the inter-packet interval to $1/r_a$ seconds to produce a smooth flow with rate r_a packets per second. The arrival process of TFRC packets at the BS buffer can be approximated by a stationary Bernoulli process, and the probability of a packet arrival in a given time slot is $r_a T$. Simulation results show that the Bernoulli model is acceptable.

When the buffer is non-empty, a packet is sent at the beginning of the slot. It is assumed that the channel condition remains at its current state during the transmission of a packet. If the transmission is successful, the packet leaves the buffer at the end of the slot.

At the beginning of each slot, let n denote the number of packets in the buffer, s the channel state, and m the number of times the packet currently being served has been transmitted. A three-dimensional DTMC is defined as (n, s, m), where $0 \le n \le B$, $s \in \{b, g\}$, and $1 \le m \le M$. This DTMC is referred to as the *queue-associated* DTMC to differentiate it from the other DTMC to be developed in Subsection 7.3.2. Obviously, states with both $n = 0$ and $m > 1$ are unreachable. The state transition diagram is shown in Figure 7.2, and the state transition probabilities are listed in Table 7.2.

Given $r_a T$ and the state transition probability matrix P, the steady state distribution, $\pi(n, s, m)$, can be derived according to the following balance equations:

$$\begin{cases} \mathbf{\Pi}^T P = \mathbf{\Pi}^T \\ \sum_{n=0}^{B} \sum_{s \in \{b,g\}} \sum_{m=1}^{M} \pi(n, s, m) = 1, \end{cases} \tag{7.2}$$

where $\Pi = [\pi(n, s, m)]$.

By Little's law, the average round-trip time of the TFRC flow can be derived as follows:

$$rtt = 2t + \frac{1}{r_a} \sum_{n=1}^{B} \sum_{s \in \{b,g\}} \sum_{m=1}^{M} n\pi(n, s, m). \tag{7.3}$$

A packet is lost when either the BS buffer overflows or the number of transmission attempts reaches M. Therefore, packet loss rate, p_l, can be calculated by

$$p_l = \sum_{s \in \{b,g\}} \sum_{m=1}^{M} \pi(B, s, m) + \sum_{n=0}^{B-1} \pi(n, b, M). \tag{7.4}$$

The link utilization, η, defined as the successful transmission rate over the link capacity, can be obtained as follows:

$$\eta = r_a(1 - p_l)/r. \tag{7.5}$$

Table 7.2 States transition probabilities of the queue-associated DTMC [146]. Reproduced by permission of ©2006 IEEE.

Current state	Next state	Transition probability
$(0, g, 1)$	$(0, g, 1)$	$(1 - r_aT)p_{gg}$
	$(0, b, 1)$	$(1 - r_aT)p_{gb}$
	$(1, g, 1)$	r_aTp_{gg}
	$(1, b, 1)$	r_aTp_{gb}
(n, g, m)	$(n - 1, g, 1)$	$(1 - r_aT)p_{gg}$
$1 \le n \le B - 1$	$(n - 1, b, 1)$	$(1 - r_aT)p_{gb}$
$1 \le m \le M$	$(n, g, 1)$	r_aTp_{gg}
	$(n, b, 1)$	r_aTp_{gb}
(B, g, m)	$(B - 1, g, 1)$	p_{gg}
$1 \le m \le M$	$(B - 1, b, 1)$	p_{gb}
$(0, b, 1)$	$(0, g, 1)$	$(1 - r_aT)p_{bg}$
	$(0, b, 1)$	$(1 - r_aT)p_{bb}$
	$(1, g, 1)$	r_aTp_{bg}
	$(1, b, 1)$	r_aTp_{bb}
(n, b, m)	$(n, g, m + 1)$	$(1 - r_aT)p_{bg}$
$1 \le n \le B - 1$	$(n, b, m + 1)$	$(1 - r_aT)p_{bb}$
$1 \le m \le M - 1$	$(n + 1, g, m + 1)$	r_aTp_{bg}
	$(n + 1, b, m + 1)$	r_aTp_{bb}
(n, b, M)	$(n - 1, g, 1)$	$(1 - r_aT)p_{bg}$
$1 \le n \le B - 1$	$(n - 1, b, 1)$	$(1 - r_aT)p_{bb}$
	$(n, g, 1)$	r_aTp_{bg}
	$(n, b, 1)$	r_aTp_{bb}
(B, b, m)	$(B, g, m + 1)$	p_{bg}
$1 \le m \le M - 1$	$(B, b, m + 1)$	p_{bb}
(B, b, M)	$(B - 1, g, 1)$	p_{bg}
	$(B - 1, b, 1)$	p_{bb}

The length of burst errors measures the number of consecutive packet losses due to transmission errors. Since i packets will be discarded consecutively due to transmission errors if the channel remains in the bad state for iM slots, the average

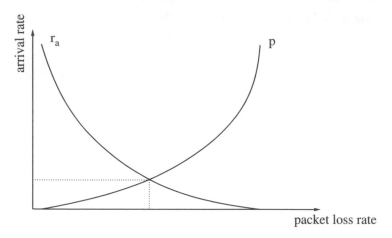

Figure 7.3 TFRC sending rate vs. packet loss rate [146]. Reproduced by permission of ©2006 IEEE.

length of burst errors in transmission is given by

$$l_b = \sum_{i=1}^{+\infty} i(p_{bb}^M)^{i-1}(1 - p_{bb}^M) = 1/(1 - p_{bb}^M). \qquad (7.6)$$

One or more packet losses during one *rtt* are counted as one loss event by the TFRC receiver. Assume that there is only one burst of losses during one *rtt* and the length of burst losses is approximated by l_b. The loss event rate can be approximated by

$$p \approx p_l/l_b. \qquad (7.7)$$

We can obtain p as a non-decreasing function of r_a according to the above analysis; on the other hand, the TFRC sender adjusts r_a as a non-increasing function of p according to (7.1). As shown in Figure 7.3, we can obtain the coordinates of the intersection point of the two curves, which correspond to the p and r_a in steady state.

The iterative algorithm to obtain p, p_l and r_a in steady state is shown in Algorithm 4 below.

Iterative computation uses binary search, which is done as follows. Set r_a^{Init} to $r/2$ initially, and calculate *rtt*, p_l and p according to (7.3), (7.4) and (7.7), respectively. Then, substitute the calculated p and *rtt* into (7.1) to obtain an updated r_a^{new}. By repeating this procedure, r_a^{new} converges to a constant, which is the steady state sending rate r_a. Thereafter, p_l can be obtained by substituting r_a into (7.2) and (7.4), and the link utilization can be obtained by substituting p_l into (7.5).

TFRC performance over a wireless link without the ARQ scheme can be obtained by setting M to 1.

Algorithm 4 Algorithm to derive r_a [146]. Reproduced by permission of ©2006 IEEE.

1: $r_a^{Init} = 0$ and $r_a^{new} = r$;
2: **while** $\left| r_a^{new} - r_a^{Init} \right| \geq \epsilon$ **do**
3:      $r_a^{Init} = (r_a^{new} + r_a^{Init})/2$;
4:      using r_a^{Init} to calculate rtt, p_l, and p according to (7.3), (7.4), and (7.7), respectively;
5:      substitute the calculated p and rtt into (7.1) to calculate r_a^{new};
6: **end while**
7: $r_a = r_a^{new}$;

7.3.2 Delay performance

To study the delay performance, we develop another DTMC as follows. When a packet arrives at the BS buffer, the queue length (including the target packet) is denoted by q. We use the triple (n_d, s_d, m_d) to represent the current state of a packet, where $1 \leq n_d \leq q$, $s_d \in \{b, g\}$ and $1 \leq m_d \leq M$. These indexes have the following significance: $n_d - 1$ is the number of packets currently being queued in the BS buffer before the target packet; s_d is the current wireless channel state; and m_d is the number of times the currently being served packet has been transmitted. The state evolves in every slot, and the sequence of the states is a *packet-associated* DTMC. For instance, if $n_d = q$ and the packet currently being served is either successfully transmitted or discarded by the BS, then $n_d = q - 1$ in the next slot; otherwise, $n_d = q$ in the next slot.

Figure 7.4 shows the state transition trajectories of the packet-associated DTMC. At the end of each trajectory, there is an absorbing state, either *(success)* or *(failure)*, corresponding to the target packet being successfully received by the MH or being discarded after M failed attempts, respectively.

The one-step state transition probabilities of the packet-associated DTMC are listed in Table 7.3, and the k-step transition matrix of the Markov chain P_k equals P_1^k. From P_k, the probability that the packet is transmitted successfully at the kth slot after its arrival, denoted by $p_{k(q,s,m),(success)}$, can be obtained, where (q, s, m) is the initial state associated with the target packet. The minimal queueing delay at the BS is $q - 1$, if the wireless channel is in the *good* state for q consecutive slots. The maximum queueing delay is $qM - 1$, if the wireless channel is in the *bad* state for $(qM - 1)$ consecutive slots and changes to the *good* state at the qMth slot. Let D be the maximum delay jitter that an application can tolerate. The conditional probability that the queueing delay of a successfully received packet is larger than D slots is given by

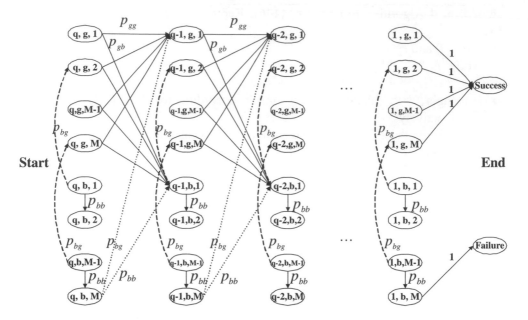

Figure 7.4   State transition diagram of the packet-associated DTMC [146]. Reproduced by permission of ©2006 IEEE.

Table 7.3 States transition probabilities of the packet-associated DTMC [146]. Reproduced by permission of ©2006 IEEE.

Current state	Next state	Transition probability
(q, g, m)	$(q - 1, g, 1)$	p_{gg}
$1 < q \leq B, 1 \leq m \leq M$	$(q - 1, b, 1)$	p_{gb}
(q, b, m)	$(q, g, m + 1)$	p_{bg}
$1 < q \leq B, 1 \leq m < M$	$(q, b, m + 1)$	p_{bb}
(q, b, M)	$(q - 1, g, 1)$	p_{bg}
$1 < q \leq B$	$(q - 1, b, 1)$	p_{bb}
$(1, g, m)$	success	1
$1 \leq m \leq M$		
$(1, b, m)$	$(1, g, m + 1)$	p_{bg}
$1 \leq m < M$	$(1, b, m + 1)$	p_{bb}
$(1, b, M)$	failure	1

$$\Pr\{t_q > D | (q, s, m)\} = \frac{\sum_{k=D+2}^{q \cdot M} p_{k(q,s,m),(success)}}{\sum_{k=q}^{q \cdot M} p_{k(q,s,m),(success)}},$$ (7.8)

where t_q represents the queueing delay at the BS buffer.

Obviously, $(1, g, 2), \ldots, (1, g, M)$ and $(1, b, 2), \ldots, (1, b, M)$ are invalid initial states, and the probability of being in one of these states initially is zero. The probability of the initial state associated with the packet can be calculated as the steady state probability of the queue-associated DTMC normalized by the sum of all possible initial states, i.e.,

$$\Pr\{(q, s, m)\} = \pi(q - 1, s, m)/\Pi_v,$$

where

$$\Pi_v = \sum_{s \in \{b,g\}} \pi(0, s, 1) + \sum_{q=2}^{B} \sum_{s \in \{b,g\}} \sum_{m=1}^{M} \pi(q - 1, s, m),$$

and $\pi(q - 1, s, m)$ has been derived in Subsection 7.3.1.

The probability of queueing delay exceeding D slots is given by

$$\Pr\{t_q > D\} = \frac{1}{\Pi_v} \left[\sum_{s \in \{b,g\}} \Pr\{t_q > D | (0, s, 1)\} \cdot \pi(0, s, 1) \right.$$

$$\left. + \sum_{q=2}^{B} \sum_{s \in \{b,g\}} \sum_{m=1}^{M} \Pr\{t_q > D | (q - 1, s, m)\} \cdot \pi(q - 1, s, m) \right]. \quad (7.9)$$

Although with a larger M the packet loss rate due to transmission errors can be reduced and the wireless link utilization can be increased, more delay and delay jitter are introduced. Similarly, with a larger BS buffer size, B, the packet loss rate due to buffer overflow can be reduced; however, it also results in more delay and delay jitter. Therefore, M and B should be chosen appropriately by making a tradeoff between system utilization and QoS requirements. For time-sensitive applications, the delay outage probability given by (7.9) should be bounded. Thus, the system parameters, M and B, can be optimized according to the wireless channel profile, to maximize the link utilization and TFRC flow throughput under the QoS constraint. This will be discussed in detail in the following section.

7.4 Performance Evaluation

To verify the analysis in Section 7.3, extensive simulations are performed using NS-2. The simulation topology is that shown in Figure 7.1.

We assume that the physical layer power control and error control schemes can adapt to transmission distance and channel conditions, such that the packet transmission error rate p_e remains constant ($p_e = 0.0952$). Although the packet error rate can be guaranteed by the physical layer protocol, with different Doppler frequency and channel coherence time, the channel state transition probabilities are different, which will affect the performance of the upper layer protocols. Since the Doppler frequency is determined by the MH velocity, the analytical and simulation results are compared with respect to the MH velocity. The numerical results can be used as guidance to select the values of M and B appropriately, to efficiently utilize the wireless links and statistically guarantee the packet loss rate and delay outage rate.

In the simulation, the following parameters are used unless otherwise explicitly stated. The downlink and uplink bandwidths allocated to the target TFRC flow are 384 Kbps and 96 Kbps, respectively. The TFRC flow has a packet size of 240 bytes, and the downlink capacity is equal to 200 packets per second. The bandwidth of the wired link between the CH and the BS is 100 Mbps. The carrier frequency of the wireless transmission is 900 MHz. The deterministic end-to-end delay t is 20 ms. The transmission time of one packet over the wireless link (the duration of a time slot) is $T = 5$ ms. The maximum tolerable delay jitter of the multimedia application is 50 ms or ten time slots. Since the transmission time of ten packets equals 50 ms, the buffer size B is set to ten packets.

Since the TFRC flow is allocated a dedicated wireless link, a simple Drop-Tail queue management scheme is deployed at the BS. The TFRC sender is saturated so that it always has data to send. Each simulation lasts for 500 seconds. We repeat the simulation with different random seeds, and obtain the simulation results by averaging those of six runs.

In this chapter, we also use the following simple approach to establish the two-state DTMC model for the wireless channel.

Given the physical layer coding and modulation schemes, when the received SNR is below a threshold, $E[SNR]/F$, the channel is in the *bad* (*b*) state; otherwise, it is in the *good* (*g*) state. $E[SNR]$ is the mean of the received SNR. As derived in [132], the average packet error rate is given by

$$p_e = 1 - e^{-1/F}. \tag{7.10}$$

Here, we use a simple method given in [132] to derive the state transition probabilities of the two-state Markov chain:

$$p_{bb} = 1 - \frac{Q(\theta, \rho\theta) - Q(\rho\theta, \theta)}{e^{1/F} - 1}, \tag{7.11}$$

$$p_{gg} = \frac{1 - p_e(2 - p_{bb})}{1 - p_e}, \tag{7.12}$$

where $\theta = \sqrt{(2/F)/(1 - \rho^2)}$ and $\rho = J_0(2\pi\mu T)$. ρ is the Gaussian correlation coefficient of two samples of the complex amplitude of a fading channel with Doppler frequency μ, sampled T seconds away. T is the packet transmission time over the wireless link, i.e., $T = 1/r$. $J_0(\cdot)$ is the Bessel function of the first kind and zero order. $Q(\cdot, \cdot)$ is the Marcum Q function.

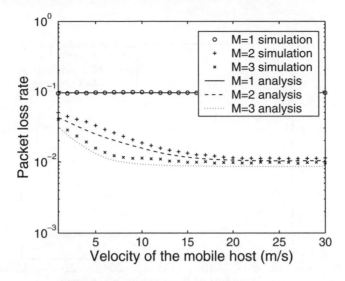

Figure 7.5 Packet loss rate, $t = 20$ ms [146]. Reproduced by permission of ©2006 IEEE.

7.4.1 Packet loss rate

Figure 7.5 compares the analytical and simulation results of packet loss rates for $M = 1, 2$ and 3.

For $M = 1$, since TFRC traffic is quite smooth and the sending rate is much less than the wireless link capacity due to the high transmission error rate, the BS queue does not build up and there is no buffer overflow. Thus, the packet loss rate is always equal to the transmission error rate p_e.

Given a *constant* transmission error rate, when the truncated ARQ is deployed, the packet loss rate decreases as v increases. This is because, with higher velocity, the Doppler frequency is higher, and the channel coherence time is smaller. Thus, the length of burst errors is shorter, and the truncated ARQ scheme can effectively recover burst errors with length no larger than M.

The average length of burst errors (l_b), determined by the wireless channel state transition probability and M, is shown in Figure 7.6. It can be seen that, with higher velocity, l_b is smaller since the fading process becomes more uncorrelated. The packets will be discarded due to transmission errors only if l_b is larger than M. Therefore, when the velocity is low, packet loss rate decreases as M increases; when the velocity is quite high and l_b approaches 1, there is no significant difference with $M = 2$ or $M = 3$. As shown in Figure 7.5, when the velocity is larger than 20 m/s, the packet loss rates with $M = 2$ and $M = 3$ are almost the same.

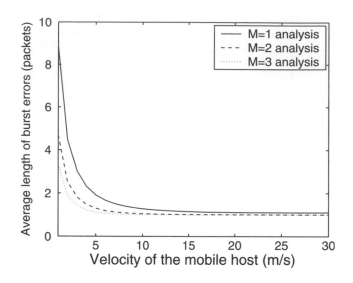

Figure 7.6   Average length of burst errors, $t = 20$ ms [146]. Reproduced by permission of ©2006 IEEE.

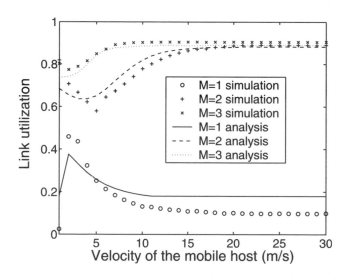

Figure 7.7 Link utilization, $t = 20$ ms [146]. Reproduced by permission of ©2006 IEEE.

7.4.2 Link utilization

Figure 7.7 shows the analytical and simulation results of wireless link utilization, which can also be viewed as normalized flow throughput.

When $M = 1$, the throughput decreases as the velocity of the mobile host v increases. In other words, without the ARQ scheme, TFRC performs better in a burst error environment than in a random error environment. The reason for this is because, according to (7.1), the TFRC sending rate is inversely proportional to the square root of loss event rate p, which is approximately p_l/l_b. Thus, at low velocity, larger l_b leads to a smaller loss event rate and higher link utilization. However, if l_b becomes so large that there is no acknowledgment received at the sender for a while, the sender timer will expire. According to the TFRC protocol specification, when timeout occurs, the sending rate is cut by half, which results in an even lower flow throughput. As shown in Figure 7.7, the link utilization increases with v when $v < 2$ m/s due to fewer timeouts; subsequently, the utilization decreases as the velocity increases due to a smaller value of l_b.

When $M = 2$ and 3, the figure shows that the bandwidth utilization is high for low velocity, decreases for medium velocity and then increases for high velocity. By comparing Figures 7.5 and 7.6 at low velocities, although the packet loss rate is high, due to large l_b, the loss event rate is small. Thus, the TFRC throughput and link utilization at a low velocity is higher than that at a medium one. When the velocity becomes even higher, l_b decreases slower than the packet loss rate; therefore, the TFRC flow throughput and the link utilization become higher.

In Figure 7.7, when $M = 2$ and $v > 20$ m/s, or $M = 3$ and $v > 10$ m/s, the link utilization can approach the upper bound, $1 - p_e$, i.e., the wireless link can be fully utilized by the TFRC flow.

7.4.3 Delay outage rate

When $M = 2$ and 3, the analytical and simulation results of delay outage rates (the ratio of packets with delay jitter t_q larger than 50 ms) are shown in Figure 7.8. For the TFRC flow, since higher sending rate (throughput) leads to higher buffer occupancy and longer queueing delay, the delay outage rate curve follows the utilization curve, as shown in Figures 7.7 and 7.8.

With $M = 1$, the queueing delay is negligible, since without the ARQ scheme, TFRC throughput is much lower than the link capacity, as shown in Figure 7.7. Therefore, the delay outage rate is zero when $M = 1$.

7.4.4 Effect of deterministic end-to-end delay

To evaluate the effect of deterministic end-to-end delay, simulation results with $t = 20$ ms and $t = 50$ ms are compared and shown in Figures 7.9 and 7.10.

In Figure 7.9, for $M = 3$, with low velocity, most of the losses are due to those packets being discarded by the BS when three attempts failed. Thus, TFRC flows

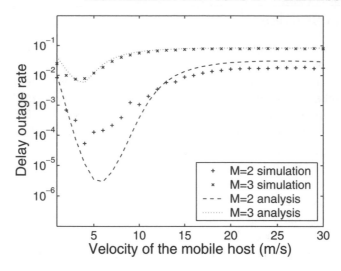

Figure 7.8 Delay outage rate, $t = 20$ ms [146]. Reproduced by permission of ©2006 IEEE.

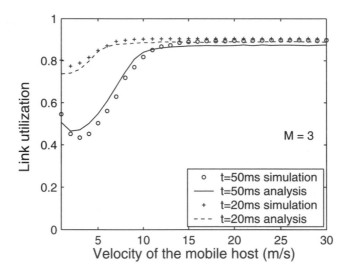

Figure 7.9 Link utilization, $t = 20$ ms (dash line) vs. $t = 50$ ms (solid line) [146]. Reproduced by permission of ©2006 IEEE.

with different propagation delays have approximately the same packet loss rate, which is determined by the wireless link statistics only. With the same loss rate, the TFRC sending rate is inversely proportional to *rtt*, and the TFRC flow has lower throughput and link utilization with larger t.

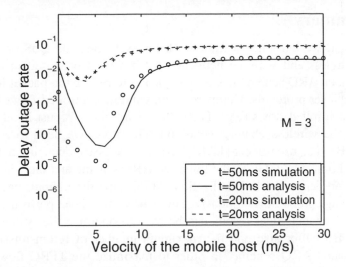

Figure 7.10 Delay outage rate, $t = 20$ ms (dash line) vs. $t = 50$ ms (solid line) [146]. Reproduced by permission of ©2006 IEEE.

With high velocity, most packets can be successfully transmitted within three attempts and packet losses are mainly due to buffer overflow. Therefore, the TFRC flow can fully utilize the wireless link and the throughput approaches the maximum value, even with a large t. In other words, given appropriate buffer size, if there is no packet loss due to transmission errors, the sending rate of TFRC flows can always approach the link capacity no matter how large t is.

As shown in Figure 7.10, the delay outage curve follows the link utilization curve, and the analytical results match well with the simulation ones.

In summary, based on the wireless channel profile (transmission error rate, state transition probability, etc.) and the QoS requirements (the maximum tolerable delay jitter, delay outage rate, packet loss rate, etc.), the BS can determine the suitable values of B and M to maximize the link utilization and the TFRC flow throughput under the QoS constraint.

For example, a multimedia application requires that the delay outage rate (the ratio of packets with delay jitter exceeding 50 ms) should be less than 1% and the packet loss rate should be less than 2%. When $t = 20$ ms and v is around 15 m/s, the BS can set $M = 2$ and $B = 10$ packets since the delay outage rate will exceed the target one if $M = 3$. In this setting, the link utilization is around 0.84, and the efficiency does not change much when setting $M = 3$. On the other hand, if the velocity is around 5 m/s, the BS can set $M = 3$ and $B = 10$ packets, so the delay outage rate and packet loss rate can be guaranteed, and the link utilization is 20% higher than that with $M = 2$.

7.5 Summary

In this chapter, we introduce a framework for evaluating the QoS of TFRC flows in hybrid wireless and wireline networks that has been proposed in [146]. In the wireless link, an ARQ scheme has been deployed to reduce the packet losses visible to the higher layer protocols. Compared with window-controlled flows (e.g., AIMD flows), rate-controlled flows (e.g., TFRC flows) are not as robust in adapting to the variation of the wireless channel condition. Therefore, we need to set tight retry limits at the BS (i.e., use truncated ARQ) to avoid excessive queueing delay when the channel condition is bad. Given the truncated ARQ scheme and the wireless channel profile, we can derive the link utilization, TFRC flow throughput, packet loss rate and delay outage probability. Extensive simulations have been performed to validate the analysis. The analytical results can be used to optimize the system parameters such as the BS buffer size and the maximum number of retransmission attempts for the truncated ARQ scheme, in order to maximize the TFRC flow throughput and wireless link utilization, and to satisfy the QoS requirements of time-sensitive multimedia applications.

7.6 Problems

1. Let a TFRC sender transmit a very large file over a wired link with capacity C. The round-trip time for the flow without queueing delay is R. No other traffic shares the link with it. Assume that there is no transmission error in the link and no packet loss in the reverse path from the TFRC receiver to the TFRC sender. The routers' buffer size B is much smaller than the file size, and the Drop-Tail queue management scheme is deployed. What is the packet loss rate of the TFRC flow in steady state?

2. Let a flow occupy a wireless link. The data from the CH is delivered to the MH through the BS which deploys truncated ARQ. Let the transmission time of a packet over the wireless link be the duration of a time slot. The data source generates Bernoulli traffic, and the probability of a packet arrival in a given time slot is a, where $a \in (0, 1)$. Assume a perfect and instantaneous wireless uplink channel so that, at the end of each time slot, the BS knows whether the transmission was successful or not, and it will retransmit the corrupted packet immediately in the following slot up to M times. Similar to the approach in Section 7.3, we can establish a queue-associated three-dimensional DTMC to investigate the system performance. Answer the following questions and give your justifications.

    (a) Draw the state transition diagram and derive the state transition probabilities of the queue-associated three-dimensional DTMC.

    (b) What are the steady state distributions of the queue-associated three-dimensional DTMC?

(c) What is the average queueing delay in the wireless buffer?

(d) What is the packet loss probability due to buffer overflow?

(e) What is the packet loss probability due to failed retransmissions?

(f) What is the link utilization?

3. Let a TFRC flow be established between a CH and an MH through a BS. The TFRC sender resides in the CH and the TFRC receiver resides in the MH. The bottleneck for the TFRC flow is the downlink between the BS and the MH, which can transmit 200 packets per second for the TFRC flow. In the absence of queueing delay, the one-way end-to-end delay is 20 ms. The maximum tolerable delay jitter of the multimedia application is 50 ms. The buffer size at the BS is set to 10 packets. The wireless link deploys truncated ARQ, and each packet will be transmitted up to M times.

(a) If there is no transmission error, what is the throughput of the TFRC flow?

(b) The wireless link is modeled as a two-state Markov chain with time slot equal to 5 ms, and the state transition probabilities are $p_{bb} = 0.5$ and $p_{gg} = 0.99$. Let $M = 3$.

 (i) What is the throughput of the TFRC flow?

 (ii) What is the link utilization?

 (iii) What is the maximum queueing delay for the TFRC packets?

 (iv) What is the delay outage rate for the TFRC flow?

(c) The wireless link has the following state transition probabilities: $p_{bb} = 0.5$ and $p_{gg} = 0.95$. How do you choose M to ensure that the delay outage rate for the TFRC flow is below 1%?

4. Compare the performance of the UDP, AIMD and TFRC protocols in support of multimedia applications with stringent delay requirements over wireless networks, and list their advantages and disadvantages.

5. Someone changes the TFRC sender's algorithm, so that the hacked TFRC sender sets the sending rate to be twice that given in (7.1). Repeat question 3 with the hacked TFRC.

6. Consider the following wireless channel and repeat question 3. The wireless link is modeled as a two-state Markov chain with time-slot duration equal to 5 ms, and the state transition probabilities are $p_{bb} = 0.5$ and $p_{gg} = 0.99$. In different channel states the transmission rates are different, and the packet error rates in both states are negligible. In the *good* state and the *bad* state, it takes 5 ms and 10 ms to transmit a packet, respectively.

8

Multimedia Services in Wireless Random Access Networks

Since the wireless communication medium is both multiple access and broadcast in nature, access control schemes are required to coordinate users in an area to share the resources. Wireless medium access can be classified into two categories: (a) centralized resource allocation and management and (b) distributed random access. With centralized resource allocation and management schemes, wireless resources (e.g., time-slots, frequency bands, or codes in a TDMA, FDMA or CDMA system) are usually allocated by a centralized controller in a deterministic manner. In a dynamic network with intermittent users and bursty traffic, pre-allocation of resources involves high protocol overheads and is hence inefficient. Therefore, in computer communications networks, random access schemes are very popular because of their simplicity, scalability and robustness.

8.1 Brief History of Random Access Technologies

The simplest random access technology is the pure Aloha MAC protocol. Pure Aloha was developed in the 1970s by Professor Norman Abramson and his associates at the University of Hawaii, when they built a radio-based data communications system to connect the Hawaiian islands, and to connect them with mainland USA. With the pure Aloha MAC protocol, a user can transmit data at any time; if collision happens, the data is sent later. As illustrated in Figure 8.1(a), a packet transmission is successful if and only if nobody else transmits during its vulnerable period. With pure Aloha, the vulnerable period is twice the packet transmission time (T). Assume that the aggregated new and retransmission arrival process is Poisson distributed with average offered traffic load G (packets per T s). A transmission can be successful if no other transmissions occur during its vulnerable period of $2T$. The probability of no arrival

Multimedia Services in Wireless Internet: Modeling and Analysis Lin Cai, Xuemin Shen and Jon W. Mark
© 2009 John Wiley & Sons, Ltd

during $2T$ is $\exp(-2G)$, which equals the probability of success. Therefore, the average network throughput (or bandwidth utilization), i.e., the number of successful transmissions during T, is $G \exp(-2G)$. The maximum bandwidth utilization of a wireless network using pure Aloha is $1/2e$, which occurs when $G = 0.5$, as shown in Figure 8.1(b).

Without any overhead of resource allocation and synchronization, pure Aloha is particularly suited for ad hoc networks with low-volume traffic or simple wireless devices. For instance, in some wireless sensor networks where the sensor nodes are equipped with transmitter only, pure Aloha is probably the best choice in the MAC layer. However, if the network traffic volume is high, collisions happen frequently, which wastes wireless bandwidth and energy.

To improve the efficiency of pure Aloha, slotted Aloha was introduced. Time is partitioned into slots, each with a duration equal to the packet transmission time T, and each user is allowed to transmit only at the beginning of a time slot. According to Figure 8.1(a), collisions occur when packets are completely overlapped, so the vulnerable period is reduced to T. Thus, the probability of successful transmission is $\exp(-G)$, the average network throughput is $G \exp(-G)$ (packets per T) and the maximum bandwidth utilization is $1/e$, which occurs at $G = 1$, as shown in Figure 8.1(b). Although slotted Aloha can double the maximum throughput, the cost of synchronization is not trivial. In addition, the maximum bandwidth utilization of $1/e$ is still not satisfactory when the wireless spectrum is at a premium.

In WLANs, carrier sense multiple access (CSMA) is used to significantly improve the resource utilization. With a CSMA MAC protocol, each user senses the channel before transmission. A user is not allowed to transmit if the channel is sensed busy. As shown in Figure 8.2, with CSMA, two transmissions will collide if and only if their starting time is within two times the propagation delay δ. Therefore, the vulnerable period of each transmission is 2δ. Since the propagation delay in a WLAN is in the order of microseconds, which is typically several orders shorter than the transmission time of a single packet, the bandwidth utilization of CSMA is much higher than that of pure Aloha or slotted Aloha.

8.2 IEEE 802.11 Protocol

The IEEE 802.11 standard defines two modes of MAC protocol for WLANs: the mandatory Distributed Coordination Function (DCF) mode and the optional Point Coordination Function (PCF) mode. Although the PCF mode is designed for real-time traffic [147, 148], it is not widely deployed due to its inefficient polling schemes, limited QoS provisioning, and implementation complexity.[1] On the other hand, supporting multimedia traffic over DCF-based WLANs poses significant challenges,

---

[1] Since both DCF and PCF have limited support for real-time applications, the IEEE 802.11e has been proposed to enhance the current 802.11 MAC to support applications with stringent QoS requirements [149], but it is unclear how successful the new standard will be for wide deployment.

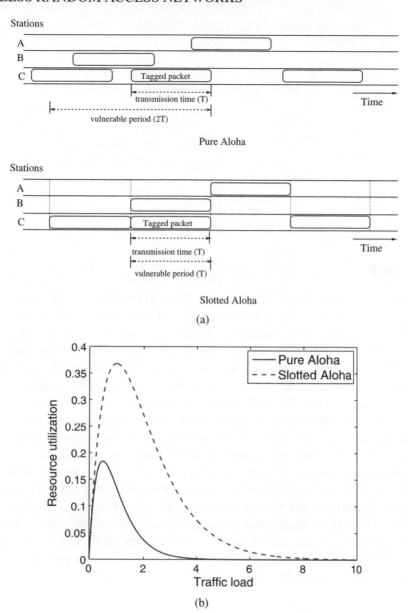

Figure 8.1 Pure Aloha vs. slotted Aloha.

since the performance characteristics of their physical and MAC layers are much worse than their wireline counterparts and other wireless systems with deterministic resource allocation schemes.

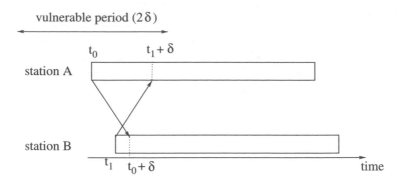

Figure 8.2 Carrier sense multiple access.

8.2.1 DCF

The IEEE 802.11 DCF-based MAC protocol uses the CSMA/CA mechanism [150], where CA stands for collision avoidance. A station monitors the medium before attempting to transmit. If the medium is sensed busy, the station defers transmission until the medium is sensed idle for a period of time equal to a DCF InterFrame Space (DIFS). After the DIFS medium idle time, it enters the backoff phase in which it sets a random backoff counter randomly chosen from [0, CW), where CW is the contention window size. The backoff counter decreases by 1 for every time slot if the medium is idle; otherwise, the counter freezes, and the decrement resumes after the medium is sensed idle again for a DIFS. When the backoff counter reaches zero, the station transmits the frame. If another station transmits a frame at the same time, a collision occurs and both transmissions fail. After an unsuccessful transmission, CW is doubled until it reaches the maximum value (CW_{max}), and the sender reschedules the transmission by randomly choosing a backoff counter in [0, CW). The frame is dropped when the retransmission limit is reached. After a successful transmission, CW is reset to its minimum value (CW_{min}). If receiving a frame successfully, the receiver transmits an acknowledgment (ACK) following a Short InterFrame Space (SIFS).

8.2.2 Ready-to-send/clear-to-send

Two medium access techniques are specified in DCF: the basic access mechanism and the ready-to-send/clear-to-send (RTS/CTS) mechanism. The RTS/CTS mechanism was designed to solve the hidden terminal problem in multi-hop wireless environments. As depicted in Figure 8.3(a), the transmission coverage and the carrier sense range of each node is assumed to be a circle centered at the node. Since the transmission of node A cannot reach node C in the shaded area, node C may not be able to sense the ongoing transmission from A and we called nodes

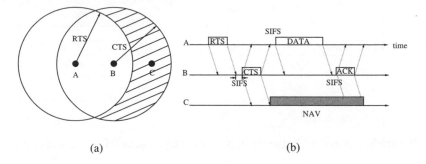

(a)   (b)

Figure 8.3 Hidden terminal problem.

in the shaded area 'hidden terminals' of node A. Without the knowledge of ongoing transmissions from node A, node C may transmit simultaneously and thus lead to a collision at the receiving node B. Similarly, node A is the hidden terminal of node C. Because the sending node is not aware of the channel condition at the receiver side, a transmission may collide with the transmissions of its neighbor(s) during the time period of 2δ, and it may collide with those of its hidden terminals during $4\delta + 2T + 2SIFS$. Therefore, the vulnerable period of a transmission in a multi-hop wireless environment is much longer due to the hidden terminal problem.

To alleviate the negative impact of the hidden terminal problem, a sending node should transmit an RTS message first to its neighbors. As shown in Figure 8.3 (b), if the receiving node finds the channel around it is idle, it can send back a CTS message after SIFS. When the handshake of RTS and CTS messages is successful, the sender can send data which is followed by an ACK from the receiver. The RTS/CTS control messages are used to reserve the channel around the sender and receiver for the following data/ACK transmissions. In the RTS message, the sender will include a Network Allocation Vector (NAV) field which indicates for how long the channel should be reserved for the following transmission. The receiver will also include the NAV field in the CTS message so the hidden terminal knows how long the channel has been reserved and it should not attempt to access the channel during that period. Using the RTS/CTS mechanism, a transmission attempt (RTS) will be collided only if there are other RTS messages transmitted by its neighbors during the 2δ time period or by its hidden terminals during the $4\delta + 2RTS + 2SIFS$ time period. If the RTS message is much smaller than the data packet, the vulnerable period with RTS/CTS is reduced and the bandwidth efficiency of the network can be improved. Thus, frames are transmitted using the RTS/CTS mechanism if their payload exceeds a given threshold; otherwise, the basic access mechanism is used.

8.3   WLAN with Saturated Stations

Using the DCF MAC protocol, a WLAN is a very dynamic system. The service time of each packet depends on the random backoff time and the collision probability,

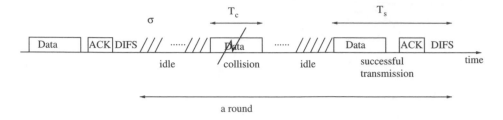

Figure 8.4 A round between two consecutive successful transmissions.

which are related to the traffic load of other competing stations. To understand the performance of DCF-based WLANs, Bianchi developed a bi-dimensional DTMC model to calculate the system throughput as a function of the number of saturated stations [151]. Here, a saturated station always has a frame ready for transmission. In this section, we first introduce Bianchi's work on how to derive the network throughput with saturated stations; then, we further study how to estimate the average service time.

8.3.1 Throughput analysis

Consider a WLAN with N saturated stations and all stations are within the transmission range of other stations. We define the time between two consecutive successful transmissions as one round, as shown in Figure 8.4. The throughput of the system (S) equals the average payload of a packet $(E[\text{Payload}])$ over the average duration of the round:

$$S = \frac{E[\text{Payload}]}{E[\text{length of a round}]}. \tag{8.1}$$

A round contains a number of idle slots when all stations are in the backoff stage, a number of collisions when two or more stations transmit simultaneously and a successful transmission period when there is only one station transmitting. Here, we assume that the transmission error is negligible so a transmission is successful if there arc no other simultaneous transmissions.

The duration of an idle slot is a constant, σ, which is specified in the IEEE 802.11 standard. Each successful transmission and each collision requires a slot of duration T_s and T_c, respectively. T_s and T_c in the basic access mode can be obtained as

$$\begin{cases} T_s^{bas} = H + E[P] + SIFS + \delta + ACK + DIFS + \delta, \\ T_c^{bas} = H + E[P^*] + DIFS + \delta, \end{cases} \tag{8.2}$$

where H is the packet header transmission time (including both PHY layer and MAC layer headers), $E[P]$ is the average transmission time of a packet payload, and $E[P^*]$ is the average transmission time of the longest packet payload involved in a collision.

Similarly, we can obtain T_s and T_c in the RTS/CTS access mode:

$$
\begin{cases}
T_s^{rts} = RTS + SIFS + \delta + CTS + SIFS + \delta + H + E[P] + SIFS + \delta \\
\qquad + ACK + DIFS + \delta, \\
T_c^{rts} = RTS + DIFS + \delta.
\end{cases}
\tag{8.3}
$$

The average number of slots in a round is

$$
E[\text{number of slots in a round}] = \frac{1}{P_{tr} P_s},
\tag{8.4}
$$

where P_{tr} is the probability that there is at least one station transmitting in a slot, and P_s is the conditional probability that a transmission is successful.

The durations of an idle slot, a transmission slot and a collision slot are different, and the average duration of each slot is

$$
E[\text{slot duration}] = (1 - P_{tr})\sigma + P_{tr} P_s T_s + P_{tr}(1 - P_s)T_c.
\tag{8.5}
$$

From (8.4) and (8.5), we can obtain the average duration of a round:

$$
E[\text{length of a round}] = \frac{(1 - P_{tr})\sigma + P_{tr} P_s T_s + P_{tr}(1 - P_s)T_c}{P_{tr} P_s}.
\tag{8.6}
$$

The key assumptions are: in steady state, each saturated station will access the channel with probability τ, the collision probability of each attempt is p, and τ and p are independent of the backoff stage of the node. For a network with N saturated stations, we have

$$
\begin{cases}
P_{tr} = 1 - (1 - \tau)^N, \\
P_s = \dfrac{N\tau(1 - \tau)^{N-1}}{P_{tr}} = \dfrac{N\tau(1 - \tau)^{N-1}}{1 - (1 - \tau)^N},
\end{cases}
\tag{8.7}
$$

and

$$
p = 1 - (1 - \tau)^{N-1}.
\tag{8.8}
$$

The remaining question is how to obtain the packet transmission probability τ and collision probability p, which are the key parameters to obtain the system throughput. To solve this problem, Bianchi built a Markov chain model from a node's point of view.

Let $b(t)$ be the stochastic process representing the backoff time counter for a given station. A discrete and integer time scale is adopted: t and $t + 1$ correspond to the beginning of two consecutive slot times, and the backoff time counter of each station decrements at the beginning of each slot time. Again, if the channel is sensed idle, the slot duration is σ; if the channel is sensed busy due to transmission or collision, the slot duration is T_s or T_c, respectively.

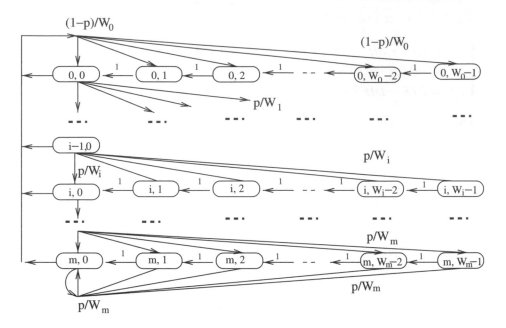

Figure 8.5 Markov chain model for the backoff window size.

Let $s(t)$ be the backoff stage of the node. Let W be $CW_{\min}$ for convenience, so that $W_i = 2^i W$ for $i \in [0, m]$. Let m, the maximum backoff stage, be the value that $2^m W = CW_{\max}$.

The node's state can be described by a bi-dimensional Markov chain, $\{s(t), b(t)\}$, and the state transition is illustrated in Figure 8.5 [151]. The state transition probabilities are

$$
\begin{cases}
P\{i, k|i, k+1\} = 1 & k \in [0, W_i - 2]i \in [0, m] \\
P\{0, k|i, 0\} = (1 - p)/W_0 & k \in [0, W_0 - 1]i \in [0, m] \\
P\{i, k|i - 1, 0\} = p/W_i & k \in [0, W_i - 1]i \in [1, m] \\
P\{m, k|m, 0\} = p/W_m & k \in [0, W_m - 1].
\end{cases}
\tag{8.9}
$$

Given the state transition probabilities and the condition that the sum of the steady state probabilities of all states should equal 1, we can obtain the steady state probabilities of all states, $b_{i,k}$ for $i \in [0, m]$ and $k \in [0, W_i - 1]$, as a function of p only. The probability τ that a node transmits in a given slot equals the sum of $b_{i,0}$ for $i \in [0, m]$:

$$
\tau = \sum_{i=0}^{m} b_{i,0} = \frac{b_{0,0}}{1 - p} = \frac{2(1 - 2p)}{(1 - 2p)(W + 1) + pW(1 - (2p)^m)}.
\tag{8.10}
$$

Equations (8.8) and (8.10) represent a nonlinear system with two unknowns, τ and p, which can be solved using numerical techniques. In addition, from (8.8),

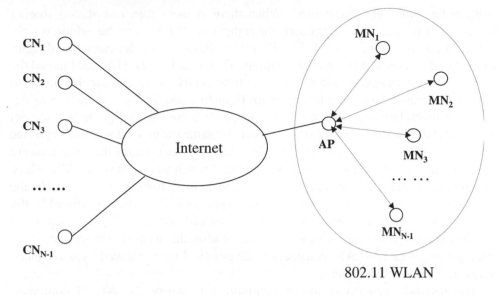

Figure 8.6 Network scenario [152]. Reproduced by permission of ©2006 IEEE.

$p(\tau)$ is a continuously and monotonically increasing function of τ for $\tau \in [0, 1]$. From (8.10), $\tau(p)$ is a continuously and monotonically decreasing function of p for $p \in [0, 1]$. Therefore, the solution of (8.8) and (8.10) is unique.

Once τ and p are obtained, they can be substituted into (8.7) and (8.6) to get P_{tr}, P_s, and E[length of a round]. Then, the system throughput can readily be obtained.

8.3.2   Average frame service time

The service time of a frame is defined as the interval from the instant the station begins to sense the channel to the instant the frame is successfully transmitted. Once we obtain the system throughput, getting the average frame service time for a single-hop WLAN with N saturated stations is straightforward. In the long term, each station has the same throughput, which equals $1/N$ of the system throughput. In other words, each station can successfully transmit one packet every N rounds. Therefore, the average service time of a frame simply equals N times E[length of a round], with the latter given by (8.6).

For delay-sensitive applications, given the average service time, we can determine the suitable buffer size according to the maximum tolerable queueing delay.

8.4   WLAN with Unbalanced Traffic

Bianchi's work assumes that all active stations are all saturated, i.e., they always have packets to send. In practice, some stations, especially those with real-time multimedia traffic, are unsaturated. When there is more than one station sharing the wireless resource, the maximum throughput of WLAN can be achieved only in the unsaturated-station case [153]. Therefore, Bianchi's model cannot be directly applied to the cases with unsaturated stations. Tickoo and Sikdar [154,155] model the queue of an unsaturated station as a discrete time G/G/1 queue, and use this model to analyze the frame delay distribution. Both Bianchi's and Tickoo's models consider a homogeneous network scenario in an independent (or ad hoc) Basic Service Set (BSS), and they assume that all stations have the same traffic load and frame service rate. However, the majority of existing WLANs are set up with the infrastructure mode, where mobile stations access the Internet through an access point (AP), which coordinates all traffic to and from the WLAN. In an infrastructure-based WLAN, the AP has a much higher traffic load and is the bottleneck. How the unbalanced traffic load affects the network performance has been investigated in [152]. The remainder of this section will introduce this work, which studies the network performance of an infrastructure-based WLAN, considering the practical issue induced by unbalanced traffic and unsaturated stations.

The network scenario is shown in Figure 8.6, where the WLAN comprises one AP and $N - 1$ mobile nodes (MNs). The AP and $N - 1$ correspondent nodes (CNs) are connected through a wired backbone network. Multimedia connections are established between MNs and CNs through the AP. By considering the different traffic load of the stations, we can obtain the conditional collision probabilities and the frame service rates of the AP and MNs. Then, we further analyze the queue utilization ratio (or traffic intensity) of the AP and MNs.

With the IEEE 802.11 DCF-based MAC, all stations have the same priority to access the channel. This is unfavorable to the AP, which has a much higher traffic load. In addition, the CSMA/CA mechanism was originally designed for data transmission, without considering delay-sensitive voice traffic. Before being successfully transmitted, each frame has to wait a random time period, which depends on the network load and collisions it has experienced. A high collision probability reduces the frame service rate and accentuates the queueing length and delay, which should be avoided for voice traffic. Thus, it is critical to obtain the upper bound of traffic load to limit the contention and collisions. We use voice traffic as an example to illustrate this problem. Given a voice codec, an accurate upper bound on the number of simultaneous voice connections that can be supported in an infrastructure-based WLAN is obtained, which will be discussed in Subsection 8.4.2.

8.4.1 Analytical model

We need to develop an analytical model suitable for investigating the performance of DCF-based WLANs with asymmetric traffic and unsaturated stations.

Consider a single-hop fully-connected WLAN with N stations, and every mobile station can sense the status of the shared wireless channel. Time is discretized into slots, and all stations operate in slotted time. Wireless channel condition is assumed ideal such that all transmitted frames can be received error-free if there is no collision. Define the conditional collision probability p_i as the probability of a collision seen by a frame being transmitted by the tagged station i. p_i is assumed constant and independent of the number of retransmissions the packet experienced. Since the probability that three or more stations simultaneously transmit is very small, we assume that a collision occurs due to two stations transmitting simultaneously.

Let the traffic arrival rate and the frame service rate of station i be denoted as λ_i and μ_i frames per slot, respectively, where $i = 0, 1, \ldots, N - 1$. The queue utilization ratio of station i is $\rho_i = \lambda_i/\mu_i$. All frames are transmitted at the same transmission rate, and the following analysis can be extended to consider rate-adaptive schemes.

Define $p_i[T]$, the probability that station i transmits a frame in a randomly chosen slot. Conditional on the queue state, the transmission probability of station i can be derived as

$$p_i[T] = p_i[T|QE]p_i[QE] + p_i[T|QNE]p_i[QNE], \qquad (8.11)$$

where $p_i[QE]$ and $p_i[QNE]$ are the probabilities of an empty queue and a nonempty queue at station i, respectively.

The queue of a station is considered empty when the station is idle, i.e., no frame is in service or waiting for service. A station is idle with probability $1 - \rho_i$ for $\rho < 1$. Therefore, $p_i[QE] = 1 - \rho_i$ and $p_i[QNE] = \rho_i$. Since a station never transmits with an empty queue, $p_i[T|QE] = 0$. Define $\tau_i = p_i[T|QNE]$ to simplify the notation, so the transmission probability of station i is

$$p_i[T] = 0 * (1 - \rho_i) + \tau_i * \rho_i = \rho_i \tau_i, \qquad (8.12)$$

where $\rho_i < 1$. Here, we only consider the unsaturated case, since, when $\rho \geq 1$, the queue is unstable and will build up, introducing excessive delay for multimedia traffic.

If station i transmits in a given slot, a collision occurs if at least one of the remaining stations also transmits in the same slot. We have

$$p_i = 1 - \prod_{j=0, j \neq i}^{N-1} (1 - p_j[T]) = 1 - \prod_{j=0, j \neq i}^{N-1} (1 - \lambda_j \tau_j/\mu_j), \qquad (8.13)$$

where $i = 0, 1, \ldots, N - 1$.

Conditional on a nonempty queue, the transmission probability of station i can be approximated as

$$\tau_i = \frac{E[M_i]}{\overline{w_i}}, \qquad (8.14)$$

where $\overline{w_i}$ is the average backoff time for station i to successfully transmit a packet and $E[M_i]$ is the average number of transmission attempts station i made during $\overline{w_i}$. Each transmission attempt has a collision probability of p_i and a success probability of $1 - p_i$. With the IEEE 802.11 DCF mode, a backoff counter is uniformly chosen over $[0, CW)$, where CW is the current contention window size. The exponential backoff procedure of a station can be modeled as a geometrically distributed random variable. Thus, the average backoff time of station i can be derived as

$$\overline{w_i} = (1 - p_i)\frac{W - 1}{2} + \cdots + p_i^{m'}(1 - p_i)\frac{\sum_{i=0}^{m'} 2^i(W - 1)}{2}$$

$$+ \cdots + p_i^m \frac{\sum_{i=0}^{m'} 2^i(W - 1) + (m - m')2^{m'}(W - 1)}{2}, \qquad (8.15)$$

where m' is the maximum backoff stage, m is the retransmission limit, and W is the minimum backoff window size. According to the IEEE 802.11 standard, m' is 5, m is 7, and W is 32 time slots. Similarly, the transmission attempts by station i can also be modeled as a geometrically distributed random variable, and the average number of transmission attempts by station i can be derived as

$$E[M_i] = (1 - p_i) \cdot 1 + \cdots + p_i^m \cdot (m + 1)$$

$$= \frac{1 - p_i^{m+1}}{1 - p_i}. \qquad (8.16)$$

Using the system parameters defined by the standard, $E[M_i]$ and $\overline{w_i}$ can be expressed as a function of p_i. Substituting (8.15) and (8.16) into (8.14), τ_i can readily be represented as a function of p_i.

To determine p_i and ρ_i, we need to obtain the average service time of a frame, $1/\mu_i$, which is the time interval between the time instant that a frame is ready to be transmitted by station i and the time instant that the frame is successfully transmitted. During $1/\mu_i$, besides a successful transmission by the tagged station i, the following events may occur:

- successful transmissions by the remaining $N - 1$ stations;

- collisions;

- channel idle when station i is in its backoff stage(s).

The transmission time of the frame sent by station i is T_{s_i}, which is the time duration the channel is sensed busy due to a successful transmission by station i. We study the system operated in a stable condition, i.e., all incoming packets are transmitted within a finite delay.

During $1/\mu_i$, on average the remaining stations successfully transmit $(1/\mu_i)\sum_{j=0, j \neq i}^{N-1} \lambda_j$ frames, which contribute $(1/\mu_i)\sum_{j=0, j \neq i}^{N-1} \lambda_j T_{s_j}$ time slots. Before the stations successfully transmit the frames, the total amount of collision time each station experiences is $(1/\mu_i)\sum_{j=0, j \neq i}^{N-1} \lambda_j \overline{T_{c_j}} + \overline{T_{c_i}}$, where $\overline{T_{c_i}}$ is the average collision time of a frame transmitted by station i. Denote by T_{c_i} the collision time station i experiences each time a collision occurs; $\overline{T_{c_i}}$ can be derived as a function of p_i:

$$\overline{T_{c_i}} = p_i(1 - p_i) \cdot T_{c_i} + \cdots + p_i^m(1 - p_i) \cdot mT_{c_i}$$

$$= \frac{p_i(1 - (m + 1)p_i^m + mp_i^{m+1})}{1 - p_i} T_{c_i}. \qquad (8.17)$$

T_{s_i} and T_{c_i} can be obtained given the frame length of station i. Since a collision is assumed to occur due to simultaneous transmissions by two stations, the duration of the channel's being busy due to collision equals half of the total amount of collision time experienced by all stations, which is $\frac{1}{2}((1/\mu_i) \sum_{j=0, j\neq i}^{N-1} \lambda_j \overline{T_{c_j}} + \overline{T_{c_i}})$. Finally, station i spends $\overline{w_i}$ in backoff stage before it successfully transmits the current frame. Therefore, we have

$$\frac{1}{\mu_i} = T_{s_i} + \frac{1}{\mu_i} \sum_{j=0, j\neq i}^{N-1} \lambda_j T_{s_j} + \frac{1}{2}\left(\frac{1}{\mu_i} \sum_{j=0, j\neq i}^{N-1} \lambda_j \overline{T_{c_j}} + \overline{T_{c_i}}\right) + \overline{w_i}, \qquad (8.18)$$

where $i = 0, 1, \ldots, N - 1$.

Given the arrival rates $\vec{\lambda} = [\lambda_0, \lambda_1, \ldots, \lambda_{N-1}]$, the equation sets (8.13) and (8.18) can be solved numerically to obtain $\vec{p} = [p_0, p_1, \ldots, p_{N-1}]$, $\vec{\mu} = [\mu_0, \mu_1, \ldots, \mu_{N-1}]$, and $\vec{\rho} = [\rho_0, \rho_1, \ldots, \rho_{N-1}]$.

8.4.2 Case study: voice capacity analysis

VoIP over WLAN (VoWLAN) has been emerging as an infrastructure to provide wireless voice service with cost efficiency. Driven by the demand from education, health-care, retail, logistics, etc., VoWLAN will experience a dramatic increase in the near future. However, supporting voice traffic over WLANs poses significant challenges since the performance characteristics of the physical and MAC layers are much worse than their wireline counterparts. In this subsection, we derive the voice capacity of a WLAN, i.e., the maximum number of voice calls that can be supported in the WLAN with guaranteed QoS, using the analytical framework given in the previous subsection.

As shown in Figure 8.6, the DCF-based WLAN comprises one AP and $N - 1$ MNs. Each MN communicates with a CN via the AP. The voice stream of station i is modeled as a CBR traffic (without the use of silence suppression) with the arrival rate of λ_i frames per slot. We assume that all MNs in the WLAN use the same voice codec so they have the same traffic load and frame service rate, $\lambda_i = \lambda_1, i = 2, \ldots, N - 1$ and $\mu_i = \mu_1, i = 1, \ldots, N - 1$. The CBR traffic model is used to derive the voice capacity, because (a) many VoIP applications do not use silence suppression, and (b) if silence suppression is used and the traffic exhibits on/off characteristics, the upper bound derived using the CBR traffic model is robust in the worst case scenario when all voice flows are in the 'on' state. Since the number of flows in a WLAN is relatively small, a tight upper bound considering the worst case scenario is desired.

In the infrastructure-based WLAN, all traffic to the MNs is transmitted via the AP, i.e., the traffic load of the AP is $N - 1$ times that of an MN. Therefore, the traffic arrival rate of the AP is $\lambda_0 = (N - 1)\lambda_1$ frames per slot. The frame service rate of the AP is denoted as μ_0 frames per slot. The queue utilization ratios at the AP and MNs are denoted by $\rho_0 = \lambda_0/\mu_0 = (N - 1)\lambda_1/\mu_0$ and $\rho_i = \lambda_1/\mu_1$, $i = 1, \ldots, N - 1$, respectively.

According to (8.13), the conditional collision probability for frames being transmitted by the AP (p_0) and that for frames being transmitted by an MN (p_1) are given by:

$$\begin{cases} p_0 = 1 - (1 - \rho_1\tau_1)^{N-1} \\ p_1 = 1 - (1 - \rho_1\tau_1)^{N-2}(1 - \rho_0\tau_0). \end{cases} \tag{8.19}$$

From (8.14), (8.15) and (8.16), we can express τ_0 and τ_1 as a function of p_0 and p_1:

$$\begin{cases} \tau_0 = E[M_0]/\overline{w_0} \\ \tau_1 = E[M_1]/\overline{w_1}, \end{cases} \tag{8.20}$$

where

$$\begin{cases} E[M_0] = \dfrac{1 - p_0^{m+1}}{1 - p_0} \\[2mm] E[M_1] = \dfrac{1 - p_1^{m+1}}{1 - p_1} \\[2mm] \overline{w_0} = (1 - p_0)\dfrac{W}{2} + \cdots + p_0^{m'}(1 - p_0)\dfrac{\sum_{i=0}^{m'} 2^i W}{2} \\[2mm] \qquad + \cdots + p_0^m \dfrac{\sum_{i=0}^{m'} 2^i W + (m - m')2^{m'} W}{2} \\[2mm] \overline{w_1} = (1 - p_1)\dfrac{W}{2} + \cdots + p_1^{m'}(1 - p_1)\dfrac{\sum_{i=0}^{m'} 2^i W}{2} \\[2mm] \qquad + \cdots + p_1^m \dfrac{\sum_{i=0}^{m'} 2^i W + (m - m')2^{m'} W}{2}. \end{cases}$$

Since all stations are using the same voice codec, all of the voice frames have the same size. Denote T_s as the time duration the channel is sensed busy because of a successful transmission, and T_c the time duration the channel is sensed busy due to failed transmissions. In the basic access mode, T_s consists of the transmission time for the voice frame, including the headers encapsulated in each layer, a SIFS, the transmission time of an ACK frame, and a DIFS:[2]

$$T_s = T_{data} + SIFS + T_{ACK} + DIFS. \tag{8.21}$$

---

[2] The propagation delay in a WLAN is usually less than one μs, and thus it is ignored here.

T_c consists of the transmission time for a voice frame, the time waiting for ACK timeout, and a DIFS:

$$T_c = T_{data} + ACK_{timeout} + DIFS. \tag{8.22}$$

The average collision time of a frame transmitted by the AP and by an MN can be derived from (8.17):

$$\begin{cases} \overline{T_{c_0}} = \dfrac{p_0[1 - (m+1)p_0^m + mp_0^{m+1}]T_c}{1 - p_0} \\[4mm] \overline{T_{c_1}} = \dfrac{p_1[1 - (m+1)p_1^m + mp_1^{m+1}]T_c}{1 - p_1}. \end{cases} \tag{8.23}$$

From the time an AP transmits a frame until the frame is transmitted successfully, the time interval $1/\mu_0$ consists of four parts: (a) on average the remaining $N - 1$ MNs successfully transmit $(N - 1)\lambda_1/\mu_0$ frames, which contribute $(N - 1)\lambda_1 T_s/\mu_0$; (b) the AP spends T_s in transmitting the current frame; (c) before the stations successfully transmit these frames, the total time that the channel is sensed busy due to failed transmissions is $[(N - 1)\lambda_1 \overline{T_{c_1}}/(2\mu_0) + \overline{T_{c_0}}/2]$; and (d) $\overline{w_0}$ is the average backoff time the AP experiences before it successfully transmits the current frame.

Similarly, the time interval $1/\mu_1$ also consists of four parts: (a) the remaining $N - 2$ MNs and the AP contribute $(N - 2)\frac{\lambda_1}{\mu_1}T_s$ and $\frac{(N-1)\lambda_1}{\mu_1}T_s$ in successful transmissions, respectively; (b) the tagged MN spends T_s s in transmitting the current frame; (c) the collision time is $\frac{1}{2}[(N - 2)\frac{\lambda_1}{\mu_1}\overline{T_{c_1}} + \overline{T_{c_1}} + \frac{(N-1)\lambda_1}{\mu_1}\overline{T_{c_0}}]$; and (d) the average backoff time of the tagged station is $\overline{w_1}$. Therefore, the average service time for the AP and the MNs are

$$\begin{cases} \dfrac{1}{\mu_0} = \left((N-1)\dfrac{\lambda_1}{\mu_0} + 1\right)T_s + \overline{w_0} \\[4mm] \qquad + \dfrac{1}{2}\left((N-1)\dfrac{\lambda_1}{\mu_0}\overline{T_{c_1}} + \overline{T_{c_0}}\right) \\[4mm] \dfrac{1}{\mu_1} = \left((N-2)\dfrac{\lambda_1}{\mu_1} + 1 + \dfrac{(N-1)\lambda_1}{\mu_2}\right)T_s + \overline{w_1} \\[4mm] \qquad + \dfrac{1}{2}\left(\left((N-2)\dfrac{\lambda_1}{\mu_1} + 1\right)\overline{T_{c_1}} + \dfrac{(N-1)\lambda_1}{\mu_1}\overline{T_{c_0}}\right). \end{cases} \tag{8.24}$$

Note that we consider only the frames that have been received successfully. The service time of frames being dropped after m failed retransmissions is not included in the above equations. In general, the frame drop probability, $p_{drop} = p_i^{m+1}$, is negligible when p_i is small and m is large. Equations (8.19) and (8.24), along with (8.20)–(8.23), can be solved numerically to obtain p_0, p_1, μ_0, μ_1, ρ_0 and ρ_1.

A station is considered stable only if its queue utilization ratio $\rho_i < 1$, i.e., the traffic arrival rate is less than the frame service rate. In an infrastructure-based

Table 8.1   Parameters of voice over 802.11 [152]. Reproduced by permission of ©2006 IEEE.

		802.11b	802.11a
	Highest channel rate	11 Mbps	54 Mbps
	Slot time	20 μs	9 μs
	SIFS	10 μs	16 μs
	DIFS	50 μs	34 μs
	CW_{min}	32	16
	CW_{max}	1024	1024
	Retry limit	7	7
T_{voice}	PLCP and preamble	192 μs	24 μs
	MAC header + FAC	24.7 μs	5 μs
	RTP/UDP/IP header	29.1 μs	6 μs
	Voice payload	(payload *8/11) μs	(payload *8/54) μs
T_{ACK}	PLCP and preamble	192 μs	24 μs
	ACK frame	10.2 μs	2.1 μs

WLAN, the AP is the bottleneck since the traffic to all MNs has to go through the AP. Therefore, the maximum number of voice connections that can be accommodated in a WLAN can be obtained under the constraint that the AP is stable, i.e., the queue utilization ratio of the AP is $\rho_0 < 1$. In addition, the number of active stations in the WLAN can also be obtained as $\sum_{i=0}^{N-1} \rho_i = \rho_0 + (N - 1)\rho_1$.

Next, we present the maximum number of VoIP connections that can be supported in a single-AP WLAN. The main parameters of the IEEE 802.11a/b and the upper-layer-header overheads of voice frames are listed in Table 8.1. Both data and ACK frames are transmitted at the highest rate.

The 802.11b standard defines the highest rate to be 11 Mbps. The values of a slot duration, DIFS and SIFS are 20 μs, 50 μs and 10 μs, respectively. Each ACK frame has 14 bytes, and it takes $14 \times 8/11 = 10.2$ μs for transmission. Each data frame has a 34 byte MAC layer overhead and a 40 byte RTP/UDP/IP-header overhead, which take $34 \times 8/11 = 24.7$ μs and $40 \times 8/11 = 29.1$ μs to transmit. In addition, it takes 192 μs to transmit the physical layer overheads consisting of a 48 μs Physical Layer Convergence Protocol (PLCP) header and a 144 μs preamble. In the 802.11a standard, the maximum data rate is 54 Mbps, approximately five times that of 802.11b. It takes 2.1 μs, 5 μs and 6 μs to transmit the ACK frame, MAC layer overhead and the RTP/UDP/IP headers, respectively. The values of a slot time, DIFS and SIFS are 9 μs, 34 μs and 16 μs, respectively; and it takes 24 μs to transmit the physical layer overhead, which is eight times smaller than that in 802.11b.

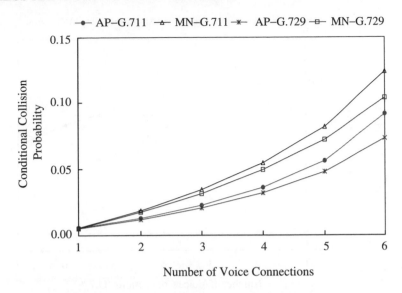

Figure 8.7 Comparison of the conditional collision probabilities of the AP and mobile nodes (802.11b) [152]. Reproduced by permission of ©2006 IEEE.

Maple 9.5 [156] is used to calculate the analytical results. Figure 8.7 shows the conditional collision probabilities of the AP and MNs with G.711 and G.729 codecs and a 10 ms packetization interval, in an IEEE 802.11b WLAN. The collisions increase with the number of voice connections. Due to the larger payload, the collision probability of G.711 is higher than that of G.729. Since the traffic load of the AP is $N-1$ times the load of an MN, collisions are more likely to occur from the viewpoint of an MN than that from the AP.

Real-time applications are very sensitive to delay and jitter. With constant arrival rate, delay guarantee of real-time applications is possible only when the traffic arrival rate is less than the service rate ($\rho_i < 1$). A station is considered unstable if its queue utilization ratio $\rho_i \geq 1$. For an unstable station, the queue will build up, thus the real-time applications will be damaged due to the ever-increasing queueing delay and packet losses due to buffer overflow.

Consider for illustration G.729 with a 10 ms packetization interval. Due to the characteristics of the voice traffic, the traffic arrival rate of an MN is constant. With an increase in the number of MNs, the traffic arrival rate of the AP increases linearly while the frame service rate exhibits a nonlinearly decreasing trend, as shown in Figure 8.8. Although the frame service rate of an MN degrades more rapidly than that of the AP due to the higher collision probability, the AP enters the unstable state before any of the MNs because of its much higher traffic load. It can be seen from Figure 8.8 that when the seventh G.729 voice connection joins in, the queue of the AP is no longer stable. Therefore, with G.729 and a 10 ms packetization interval, at most

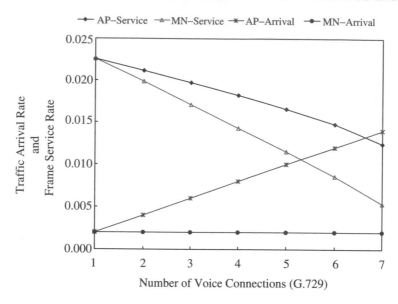

Figure 8.8 Traffic arrival rate and frame service rate (802.11b) [152]. Reproduced by permission of ©2006 IEEE.

six bi-directional VoIP connections can be supported in an IEEE 802.11b WLAN. One more VoIP connection will jeopardize the performance of all voice connections. Therefore, an accurate upper bound is critical for VoIP admission control in order to maintain an acceptable QoS for all VoIP connections.

As shown in Figure 8.9, the queue utilization ratio of the AP (ρ_0) is always much higher than that of an MN (ρ_1) due to the higher traffic load. The maximum number of voice connections with $\rho_0 < 1$ can also be observed. With the G.729 codec, 6 voice connections with a 10 ms packetization interval, 13 connections with a 20 ms interval, and 19 with a 30 ms interval can be supported in an 802.11b DCF-based WLAN. It can be seen that using G.729 or G.723 makes little difference to the maximum number of voice connections being supported in the WLAN. With G.729 or G.723, up to 19 simultaneous voice connections with a 30 ms packetization interval can be supported. The payload of G.711 is eight times that of G.729, but only two fewer connections can be accommodated. Compared with the huge overheads specified in the physical and MAC layers, the payload difference between different codecs is relatively small. The maximum number of connections with iLBC is similar to that with G.723 and G.729. For VoWLAN, G.729 and iLBC are preferred to G.723 because less compression is required.

Another observation is that more VoIP connections can be accommodated when the packetization interval is enlarged. However, a larger packetization interval will result in longer delay. There is a tradeoff between the delay constraint and the voice

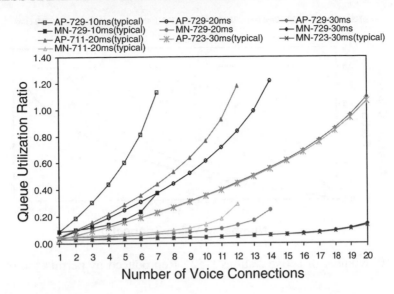

Figure 8.9 Queue utilization ratio of the AP and mobile nodes (802.11b) [152]. Reproduced by permission of ©2006 IEEE.

Table 8.2 Comparison of the maximum number of VoIP connections (802.11b) [152]. Reproduced by permission of ©2006 IEEE.

Audio (ms)	G.711			G.729			G.723			iLBC
	Proposed analysis	[158]	[157]	Proposed analysis	[158]	[157]	Proposed analysis	[158]	[157]	Proposed analysis
10	6	6	6	6	7	7				
20	11	12	12	13	14	14				12
30	15	17	18	19	21	22	19	21	22	18
40	19	21	22	25	28	28				
50	22	25	26	31	34	35				
60	25	28	29	37	41	42	37	42	42	

capacity. In addition, when we use the short preamble of 72 bits instead of the long preamble of 144 bits, two more G.711 voice connections can be admitted into the network, which indicates the significant effect of the physical layer overhead.

Table 8.2 tabulates the maximum number of VoIP connections for different codecs in an 802.11b DCF-based WLAN. It shows that only a very limited number of voice connections can be supported in a WLAN, even with the most efficient codec G.723. When the packetization interval is enlarged to accommodate more voice connections, analytical results given in [157,158] become too optimistic. This is because in [157] it is assumed that any transmitted frame is received successfully without any collision. This assumption may not hold, especially when the number of voice connections is

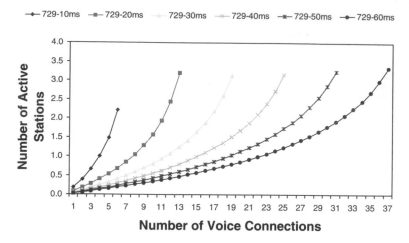

Figure 8.10 Number of active stations [152]. Reproduced by permission of ©2006 IEEE.

close to capacity. A simple approximation is made in [158]: that there are always two active stations (one is the AP and the other is an MN) in the network and the collision probability is kept as low as 0.03, independent of the number of voice connections. However, in the unsaturated-station scenario, the number of active stations is not a constant, but increases with the number and the traffic intensity of stations in the network. The AP has a frame in service with probability ρ_0 while each MN has a frame in service with probability ρ_1. On average, there are $\rho_0 + (N-1)\rho_1$ stations that have a frame in service. As shown in Figure 8.10, the average number of active stations in the WLAN varies from 0.02 when there is only one voice connection to above 3 when the AP is nearly saturated.

The data rate of an 802.11a-based WLAN is roughly five times that of an 802.11b-based one. However, the voice capacity of 802.11a is less than five times that of 802.11b due to the different parameter values specified in the standard, including the minimum contention window, duration of a slot, DIFS, SIFS, etc. For example, a smaller minimum contention window may result in more collisions and a larger SIFS causes longer service time, so both of them may reduce the voice capacity. On the other hand, smaller physical-layer overhead and shorter slot duration result in higher voice capacity. The effect of different codecs and packetization intervals on the voice capacity of an 802.11a-based WLAN is given in Table 8.3.

8.4.3   Simulation results

The analytical results are validated by extensive simulations using NS-2 (Version 2.27). The same parameter values of the IEEE 802.11b are used and listed in Table 8.1. The 802.11 code in NS-2 is modified according to the standard: the ACK

Table 8.3 Comparison of the maximum number of VoIP connections (802.11a) [152]. Reproduced by permission of ©2006 IEEE.

Audio (ms)	G.711		G.729		G.723		iLBC
	Proposed analysis	[158]	Proposed analysis	[158]	Proposed analysis	[158]	Proposed analysis
10	25	30	27	32			
20	47	56	53	64			53
30	66	79	79	95	80	96	78
40	82	98	105	126			
50	97	116	130	156			
60	110	131	155	185	158	187	

transmission rate is set to 11 Mbps and the preamble transmission rate is kept at 1 Mbps. The network topology is shown in Figure 8.6. In the wired network, the links connecting the AP and the CNs have a data rate of 100 Mbps with 20 ms propagation delay.

The end-to-end packet delay bound is set to 150 ms to maintain good voice quality.[3] Any packets that arrive after 150 ms will be discarded from the receiver's playout buffer. In order to show the queue accumulating effect in the AP, the buffer size of the AP is set to 300 packets. Initially, a voice connection is established every 10 ms (on average) to gradually approach the network capacity, with the starting time randomly chosen from [0, 10] ms. To eliminate the warming-up effects, the simulation data is collected from 10 s to 100 s.

For G.729 with a 10 ms packetization interval, the packet delay of the uplink (from an MN to the AP) and downlink (from the AP to an MN) voice flow is very low when there are fewer than six connections in the WLAN. When the seventh station joins the system, the delay of the downlink flow increases rapidly while the delay of the uplink is as low as 2 ms. This implies that the AP is saturated when the queue utilization ratio $\rho_0 \geq 1$. Meanwhile, the queue utilization ratio of the MNs, ρ_1, is much less than 1. When more stations join, which results in more collisions in the network and decreases the frame service rate, the delay of the uplink flow also increases, to more than 300 ms, implying that the MNs become saturated when there are more than 12 voice connections, as shown in Figure 8.11.

Since the downlink transmissions always suffer a longer queueing delay at the AP than the uplink transmissions at the MNs, we are more interested in the delay of downlink flows due to this bottleneck effect. Figure 8.12 shows the delay outage ratio (the ratio of the packets with end-to-end delay exceeding 150 ms over the packets being transmitted) of downlink flows, with G.729 and different packetization

---

[3]The International Telecommunication Union (ITU) has recommended one-way end-to-end delay no greater than 150 ms for good quality voice calls, with a limit of 400 ms for acceptable voice calls [159].

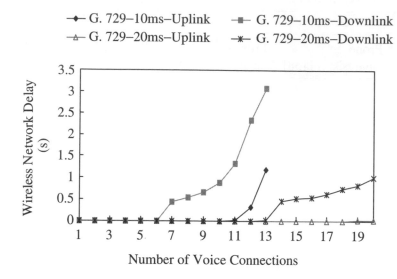

Figure 8.11 Delay comparison between uplink and downlink voice flows [152]. Reproduced by permission of ©2006 IEEE.

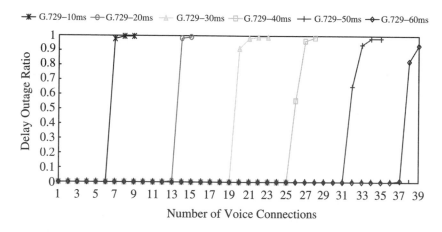

Figure 8.12 Delay outage ratio of voice traffic [152]. Reproduced by permission of ©2006 IEEE.

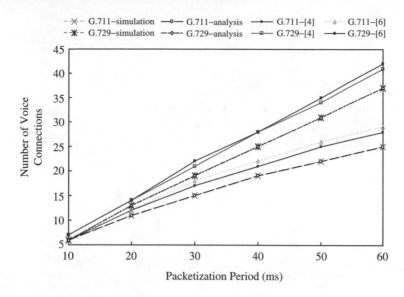

Figure 8.13   Maximum number of voice connections [152]. Reproduced by permission of ©2006 IEEE.

intervals. Due to the non-bursty characteristics of voice traffic, packet delay is quite low and no packet is discarded from the playout buffer when all stations are not saturated. However, the outage ratio of downlink flows becomes significant when the AP is saturated, due to the ever-increasing queueing delay at the AP.

In the simulation, the maximum number of voice connections is obtained in such a way that one more connection will result in a delay outage ratio larger than 1%. As shown in Figure 8.13, the analytical results approximate the simulation results quite well, and the obtained upper bounds are more accurate than the results in [157, 158], since the different collision probabilities and queue states of the AP and the MNs are considered. Therefore, the analytical results can be used as a guideline for efficient admission control.

Last, from the simulation results, all frames are transmitted within five (re)-transmissions and none is dropped by the MAC due to an excessive number of retransmissions, which validates the assumption in Subsection 8.4.1.

8.5   TFRC in the Mobile Hotspot

With the advances in wireless access technologies and the ever-increasing demand for anywhere, anytime Internet services, WLAN-based hotspot services in public areas (e.g., convention centers, cafes, airports, shopping malls etc.) are being proliferated. In addition, the extension of hotspot services to moving vehicles (e.g., subways, buses, trains, vessels, airplanes etc.) is gaining more attention [160, 161]. The hotspot

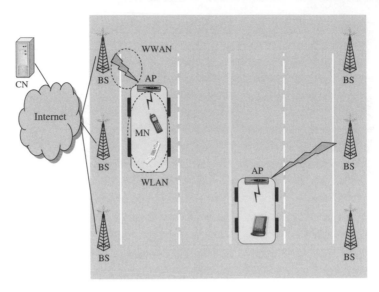

Figure 8.14 Mobile hotspot architecture [164]. Reproduced by permission of ©2008 IEEE.

service in a mobile platform is referred to as the *mobile hotspot* [162], which is a novel concept to provide ubiquitous and always best connected (ABC) services in future wireless/mobile networks.

Figure 8.14 shows a typical network architecture for mobile hotspots. Within a moving vehicle, a WLAN is used to connect a number of MNs to an AP. Meanwhile, a WWAN is employed for the connection between the AP and the BS, which is in turn connected to the Internet through a wireline link. The WWAN can be IP-based cellular systems or IEEE 802.16 WiMAX networks [163]. Packets sent from a CN to an MN in the mobile hotspot are first routed to the BS through the Internet, and then transmitted to the MN over an integrated WWAN–WLAN link. This mobile hotspot architecture has the following advantages. First, the aggregated traffic at the AP is transmitted to the BS through an antenna mounted on top of the vehicle, which has better communication channels to the BS compared with a channel between the BS and MNs inside the vehicle. Second, the AP has less energy constraint than the MNs. Using the AP to relay the data to the BS far away can significantly save the energy consumption of MNs. Third, the AP has better knowledge of the mobility and location of the vehicle, and therefore handoff management of aggregated traffic at the AP can be simpler.

Multimedia streaming is expected to be a promising application in mobile hotspots [161]. Since multimedia streaming traffic is normally long-lived and requires high data rate, flow and congestion control is important for both network stability and users' perceived QoS. In addition, fairness among multimedia flows and the currently dominant TCP controlled flows should be considered in mobile hotspots.

As discussed in the previous chapters, TFRC is a representative rate control protocol with TCP-friendliness and it is adopted here for supporting multimedia applications in mobile hotspots [143]. There are great challenges in understanding end-to-end performance of TFRC flows over mobile hotspots, due to the interaction of the end-to-end congestion control protocol and the service time variations resulting from transmission errors in the WWAN and the contentions in the WLAN.

The performance of multimedia streaming applications over heterogeneous wireless links was studied in [164]. The throughput of TFRC in mobile hotspots was investigated by developing discrete-time queueing models for the WWAN link and the WLAN link. Based on the queueing models, the packet loss probability and *rtt* of a TFRC flow were derived, and the TFRC throughput in steady state was obtained using an iterative algorithm. This study further investigated the effect on end-to-end TFRC throughput by the number of MNs in the mobile hotspot, the vehicle velocity, the WWAN/WLAN link bandwidth, the retransmission limit, and the buffer size used.

The remainder of this section will introduce the work in [164]. Subsection 8.5.1 describes the system model for the WWAN and WLAN links. In Subsection 8.5.2, the throughput model for the TFRC flow in mobile hotspots is developed and an iterative algorithm for calculating the steady state TFRC throughput is presented. Various analytical and simulation results are given in Subsection 8.5.3.

8.5.1 System description

Figure 8.15 shows the system model for analyzing TFRC throughput in mobile hotspots. We consider a TFRC flow between a TFRC sender node (SN) and a receiver node (RN). The multimedia flow is controlled by TFRC, and we focus on the downlink transmission (i.e., from the SN to the RN) for multimedia streaming applications. The model can be extended for interactive multimedia applications. Since the link layer fragmentation is not considered here, the term *packet* is used for a protocol data unit (PDU) in the data link layer or a PDU in the transport layer.

With an emphasis on packet losses in the wireless links, it is assumed that packet losses in the wired domain are negligible and the transmission delay, t_{wired}, in the wired domain between the SN and the BS is constant. For downlink transmissions in the WWAN (i.e., from the BS to the AP), a dedicated channel is allocated, so transmission errors are mainly due to channel fading. In addition, a truncated ARQ scheme is used for reliable transmission and an instantaneous feedback channel is assumed. Hence, the feedback reception time for the ARQ scheme is zero. The WWAN uplink transmissions (i.e., from the AP to the BS) of TFRC acknowledgments (*acks*) are assumed to be error-free. This assumption can be justified because the size of *acks* is relatively small and an occasional loss of *ack* can be recovered by the subsequent *acks*.

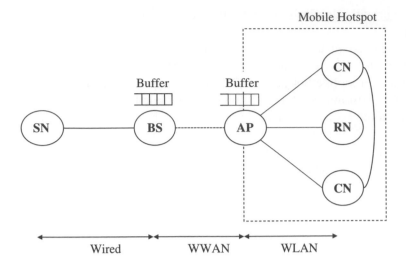

Figure 8.15 System model of TFRC in mobile hotspot [164]. Reproduced by permission of ©2008 IEEE.

The infrastructure-based WLAN is shared by N MNs, i.e., the RN and $N-1$ contending nodes, and an AP which is connected to the BS through the WWAN link. In the infrastructure-based WLAN, all traffic from and to all MNs will go through the AP. In general, the AP is the bottleneck and the MNs are unsaturated, i.e., they do not always have data to send. Since the WLAN is used for the internal connection within a vehicle, the transmission failures can be assumed due to collisions only and a collision is caused by simultaneous transmissions of two MNs or the AP and an MN.[4] The WLAN operates with the IEEE 802.11 DCF [150] in a basic access mode, since the RTS/CTS exchange is not very useful for single-hop WLANs, and is disabled in most products available in the current market [165].

Since the load of upstream traffic (from the MNs to the BS) is typically much less than that of downstream traffic (from the BS to the MNs) in multimedia streaming applications, the downlink from the BS to the AP and that from the AP to MNs are bottlenecks. The BS and the AP have buffers of sizes $B-1$ and $Q-1$, respectively.

In the following subsections, two discrete-time queueing models are developed for the WWAN and WLAN links. Table 8.4 summarizes the notation for the queueing models.

WWAN link model

For the downlink transmission from the BS to the AP, a NLOS frequency-nonselective (flat) multipath fading channel with packet transmission rate (in

[4]The collision probability due to simultaneous transmissions of three or more nodes is negligible.

Table 8.4   Notation for Section 8.5.

Symbol	Description
λ_M	Downstream transmission rate to an MN
λ_M^U	Upstream transmission rate from an MN
μ_M	Service rate at the MN queue
λ_B, λ_A	traffic arrival rate at the BS downlink and at the AP, respectively
μ_B, μ_A	Service rate of the BS downlink and of the AP, respectively
ε_W	Packet loss probability in the WWAN link
$\varepsilon_L^A, \varepsilon_L^M$	Packet loss probability in the WLAN downlink and uplink transmissions, respectively
P_B, P_Q	Blocking probability at the BS downlink and the AP downlink, respectively
p_A, p_M	Conditional collision probability for transmission by the AP and by an MN, respectively
B	AP buffer size
Q	BS buffer size
$\overline{C_A}, \overline{C_M}$	Average number of collisions during a transmission by the AP and by an MN, respectively
$\overline{BO_A}, \overline{BO_M}$	Average number of backoff slots during a transmission by the AP and by an MN, respectively
T_c	Number of time slots for a collided transmission
T_s	Number of time slots for a successful transmission
m	The maximum number of retransmissions
m'	The number of contention window sizes
θ_S	Average service time for a successful WWAN downlink transmission (in time slots)
θ_S^U	Average service time for a successful WWAN uplink transmission (in seconds)
η_S	Average service time for a successful WLAN downlink transmission (in time slots)
η_S^U	Average service time for a successful WLAN uplink transmission (in time slots)
Q_B, Q_A	Queueing delay at the BS downlink and at the AP downlink, respectively (in time slots)
σ_A, σ_M	The probability that the AP and the MN queue is nonempty, respectively
τ_A, τ_M	The probability that the AP and an MN with a nonempty queue transmits a packet in a slot

packets/seconds) much higher than the maximum Doppler frequency (in Hz) is considered. The discrete-time two-state Markov channel model is used to approximate the error process at the packet level [132]: the channel has a *good* state and a *bad* state, and packet loss probability is 1 or zero in the *bad* or *good* state, respectively. The duration of a time slot, D, corresponds to the packet transmission time over the WWAN link. Let v and f_c be the velocity of the vehicle and the carrier frequency, respectively. The Doppler frequency f_d is given by $f_c v / v_c$, where v_c is the speed of light. Let F be the fading margin. Then the average transmission error probability is $\pi_e = 1 - e^{-1/F}$ [132].

Let $\mathbf{P} = \begin{bmatrix} p_{bb} & p_{bg} \\ p_{gb} & p_{gg} \end{bmatrix}$ be the WWAN link state transition matrix. The state transition probabilities can be obtained using (7.11) and (7.12).

In the truncated ARQ scheme, if a sender experiences a transmission failure due to a bad link condition, it retransmits the packet at the next time slot. A packet is removed from the BS buffer after being successfully delivered or after l attempts (including the first transmission). Therefore, the packet loss probability in the WWAN link can be computed as

$$\varepsilon_W = \pi_e p_{bb}^{l-1}. \tag{8.25}$$

The arrival process of TFRC packets at the BS buffer can be approximated by a stationary Bernoulli process. Therefore, the WWAN downlink transmission at the BS buffer is modeled as a discrete-time $M/M/1/K$ queue with the time slot length D. Let λ_T (packets/s) be the sum of downlink transmission rates of the N MNs in a mobile hotspot. Then, the arrival rate to the BS buffer in a given time slot, λ_B (packets/slot), is computed as

$$\lambda_B = \lambda_T D. \tag{8.26}$$

On the other hand, the service rate, μ_B (packets/slot), of the BS buffer can be obtained by deriving the average service time as

$$\frac{1}{\mu_B} = (1 - \pi_e)(p_{gg} + 2 p_{gb} p_{bg} + 3 p_{gb} p_{bb} p_{bg}$$

$$+ \cdots + (l-1) p_{gb} p_{bb}^{l-3} p_{bg} + l p_{gb} p_{bb}^{l-2})$$

$$+ \pi_e (p_{bg} + 2 p_{bb} p_{bg} + 3 p_{bb} p_{bb} p_{bg}$$

$$+ \cdots + (l-1) p_{bb} p_{bb}^{l-3} p_{bg} + l p_{bb} p_{bb}^{l-2})$$

$$= (1 - \pi_e)\left(1 + \frac{p_{gb}(1 - p_{bb}^{l-1})}{1 - p_{bb}}\right) + \pi_e \left(\frac{1 - p_{bb}^{l-1}}{1 - p_{bb}} + p_{bb}^{l-1}\right), \tag{8.27}$$

where $i p_{gb} p_{bb}^{i-2} p_{bg}$ is the probability that a packet transmission is successful at the ith transmission attempt ($i > 2$) when the WWAN link state at the last transmission of the previous packet is g, whereas $i p_{bb}^{i-1} p_{bg}$ is the probability that a packet transmission is successful at the ith transmission attempt ($i > 2$) when the WWAN link state at the last transmission of the previous packet is b. Hence, the first and second terms on the right-hand side of (8.27) represent the average service times when the previous packet departs successfully and unsuccessfully, respectively.

WLAN link model

Here, we extend the queueing model in the previous section for an infrastructure-based WLAN. The downlink transmission at the AP and the uplink transmission at the MN are modeled as a discrete-time $M/M/1/K$ queue and a discrete-time $M/M/1$ queue, respectively [166], where a slot length corresponds to the length of a backoff slot δ. In IEEE 802.11 DCF, a packet is dropped if the packet transmission fails after $m+1$ attempts. Therefore, for WLAN downlink transmissions from the AP, the packet loss probability due to collisions is given by

$$\varepsilon_L^A = p_A^{m+1}, \tag{8.28}$$

where p_A is the collision probability when the AP transmits a packet. Similarly, for WLAN uplink transmissions from unsaturated MNs, the packet loss probability due to collisions, ε_L^M, can be computed as p_M^{m+1}, where p_M is the collision probability when an MN transmits a packet.

For downstream traffic, the arrival rate at the AP, λ_A (packet/slot), equals the departure rate of the successfully transmitted packets over the WWAN downlink. Then λ_A can be approximated as

$$\lambda_A = (1 - P_B)(1 - \varepsilon_W)\lambda_T\delta, \tag{8.29}$$

where P_B is the blocking probability at the BS downlink buffer. On the other hand, the service rate at the AP downlink, μ_A, can be obtained from the average service time at the AP, which includes the time for backoffs and that for successful or collided transmissions by the AP. In addition, during the interval between two packets serviced by the AP, MNs may transmit some packets successfully or experience collisions. We assume that a collision occurs mainly because of simultaneous transmissions by two stations (i.e., the AP and an MN or two MNs). Therefore, the average number of collisions during the AP service time (i.e., $1/\mu_A$) is one half of the total number of collisions during the corresponding period because a collision can be counted by two stations independently. Let λ_M^U be the upstream transmission rate of an MN. Consequently, the average service time of the AP is given by

$$\frac{1}{\mu_A} = \left(N\frac{\lambda_M^U}{\mu_A}(1 - \varepsilon_L^M) + (1 - \varepsilon_L^A) \right) T_s + \frac{1}{2}\left(N\frac{\lambda_M^U}{\mu_A}\overline{C_M} + \overline{C_A} \right) T_C + \overline{BO_A}, \tag{8.30}$$

where $(N(\lambda_M^U/\mu_A)(1 - \varepsilon_L^M) + (1 - \varepsilon_L^A))$ is the average number of successful transmissions during $1/\mu_A$ and $(N(\lambda_M^U/\mu_A)\overline{C_M} + \overline{C_A})$ is the total number of collisions during $1/\mu_A$. T_s and T_c are the numbers of time slots for successful and collided transmissions, respectively. $\overline{BO_A}$ is the average number of backoff slots during a packet transmission by the AP, and $\overline{C_A}$ and $\overline{C_M}$ are the average numbers of collisions during a packet transmission by the AP and the MN, respectively.

The probability of i collisions $(0 \le i \le m)$ for a successful transmission by the AP is $p_A{}^i(1 - p_A)$. For a failed AP transmission, $m + 1$ collisions occur and its probability is $p_A{}^{m+1}$. Then, $\overline{C_A}$ can be computed as

$$\overline{C_A} = 1 \cdot p_A(1 - p_A) + 2 \cdot p_A{}^2(1 - p_A) + \cdots + m \cdot p_A{}^m(1 - p_A)$$
$$+ (m + 1) \cdot p_A{}^{m+1}$$
$$= \frac{p_A(1 - p_A{}^{m+1})}{1 - p_A}. \tag{8.31}$$

Similarly, $\overline{C_M}$ can be computed as

$$\overline{C_M} = 1 \cdot p_M(1 - p_M) + 2 \cdot p_M{}^2(1 - p_M) + \cdots + m \cdot p_M{}^m(1 - p_M)$$
$$+ (m + 1) \cdot p_M{}^{m+1}$$
$$= \frac{p_M(1 - p_M{}^{m+1})}{1 - p_M}. \tag{8.32}$$

In the AP transmission, the probability that the ith backoff $(0 \le i \le m)$ is triggered is $p_A{}^i$. Therefore, $\overline{BO_A}$ is given by

$$\overline{BO_A} = \frac{W_0 - 1}{2} \cdot 1 + \frac{W_1 - 1}{2} \cdot p_A + \cdots + \frac{W_m - 1}{2} \cdot p_A{}^m = \sum_{i=0}^{m} \frac{W_i - 1}{2} p_A{}^i \tag{8.33}$$

where $W_i = 2^i W$ and W is the minimum contention window size. The collision and transmission time, T_c and T_s, are given by (8.2).

The service rate, μ_M, of the MN queue can be obtained as follows. During the service time of the MN, the average number of successful transmissions by other MNs except the tagged MN is $((N - 1)\lambda_M^U/\mu_M)(1 - \varepsilon_L^M)$ and the average number of successful transmissions by the AP is $[\lambda_A(1 - P_Q)/\mu_M](1 - \varepsilon_L^A)$, where P_Q is the blocking probability at the AP downlink buffer. On the other hand, the numbers of collisions by other MNs and the AP are $(N - 1)(\lambda_M^U/\mu_M)\overline{C_M}$ and $(\lambda_A(1 - P_Q)/\mu_M)\overline{C_A}$, respectively. Therefore, the average service time at the MN queue can be computed as

$$\frac{1}{\mu_M} = \left(\left((N - 1)\frac{\lambda_M^U}{\mu_M} + 1 \right)(1 - \varepsilon_L^M) + \frac{\lambda_A(1 - P_Q)}{\mu_M}(1 - \varepsilon_L^A) \right) T_s$$
$$+ \frac{1}{2} \left(\left((N - 1)\frac{\lambda_M^U}{\mu_M} + 1 \right)\overline{C_M} + \frac{\lambda_A(1 - P_Q)}{\mu_M}\overline{C_A} \right) T_c + \overline{BO_M}. \tag{8.34}$$

Similar to $\overline{BO_A}$, the average number of backoff slots with respect to the MN can be computed as

$$\overline{BO_M} = \frac{W_0 - 1}{2} \cdot 1 + \frac{W_1 - 1}{2} \cdot p_M + \cdots + \frac{W_m - 1}{2} \cdot p_M{}^m = \sum_{i=0}^{m} \frac{W_i - 1}{2} p_M{}^i.$$

(8.35)

In the unsaturated infrastructure-based WLAN, the AP and the MN transmit a packet only when their queues are non-empty. Let σ_A and σ_M be the probabilities that the AP and the MN queues are not empty, respectively. They are given by

$$\sigma_A = \frac{\rho_A(1 - \rho_A{}^Q)}{(1 - \rho_A{}^{Q+1})} \quad \text{and} \quad \sigma_M = \rho_M$$

where $\rho_A = \lambda_A / \mu_A$ and $\rho_M = \lambda_M^U / \mu_M$. By [167], the probability that the AP with a nonempty queue transmits a packet in a randomly chosen slot is given by

$$\tau_A = \begin{cases} \dfrac{2(1-2p_A)(1-p_A{}^{m+1})}{W(1-(2p_A)^{m+1})(1-p_A)+(1-2p_A)(1-p_A{}^{m+1})} & m \le m' \\[3mm] \dfrac{2(1-2p_A)(1-p_A{}^{m+1})}{W(1-(2p_A)^{m'+1})(1-p_A)+(1-2p_A)(1-p_A{}^{m+1})+W2^{m'}p_A{}^{m'+1}(1-2p_A)(1-p_A{}^{m-m'})} & m > m' \end{cases}$$

(8.36)

where m is the retransmission limit and m' is the number of contention window sizes (i.e., the maximum contention window size is $2^{m'}$). Similarly, the probability of a packet transmission by a nonempty MN is

$$\tau_M = \begin{cases} \dfrac{2(1-2p_M)(1-p_M{}^{m+1})}{W(1-(2p_M)^{m+1})(1-p_M)+(1-2p_M)(1-p_M{}^{m+1})} & m \le m' \\[3mm] \dfrac{2(1-2p_M)(1-p_M{}^{m+1})}{W(1-(2p_M)^{m'+1})(1-p_M)+(1-2p_M)(1-p_M{}^{m+1})+W2^{m'}p_M{}^{m'+1}(1-2p_M)(1-p_M{}^{m-m'})} & m > m'. \end{cases}$$

(8.37)

Then, the probabilities that the AP and the MN transmit a packet in a randomly chosen slot are $\sigma_A \tau_A$ and $\sigma_M \tau_M$, respectively. p_A and p_M are given by

$$p_A = 1 - (1 - \sigma_M \tau_M)^N$$

(8.38)

and

$$p_M = 1 - (1 - \sigma_M \tau_M)^{N-1}(1 - \sigma_A \tau_A).$$

(8.39)

Finally, (8.30), (8.34), (8.38) and (8.39) can be solved numerically to obtain μ_A, μ_M, p_A, and p_M.

8.5.2 TFRC throughput analysis

As stated in Chapter 7, the TFRC sender sets the transmission rate λ (packets/s) as [144] $1/[rtt\sqrt{2p/3} + 3p(T_0(1 + 32p^2)\sqrt{3p/8})]$, where rtt is the round-trip time, T_0 is the retransmission timeout value and p is the packet loss event rate. Therefore, we need to determine λ, rtt and p to quantify the throughput of the TFRC flow.

First, a packet in the TFRC flow can be lost due to overflows at the BS/AP buffer or transmission errors in the WWAN/WLAN link. Therefore, the packet loss rate can be obtained from

$$p = 1 - (1 - P_B)(1 - P_Q)(1 - \varepsilon_W)(1 - \varepsilon_L^A), \tag{8.40}$$

where ε_W and ε_L^A have been obtained in Subsection 8.5.1. According to queueing theory, P_B and P_Q can be computed by

$$P_B = \frac{(1 - \rho_B)\rho_B{}^B}{1 - \rho_B{}^{B+1}} \quad \text{and} \quad P_Q = \frac{(1 - \rho_A)\rho_A{}^Q}{1 - \rho_A{}^{Q+1}}, \tag{8.41}$$

where $\rho_B = \lambda_B/\mu_B$ and $\rho_A = \lambda_A/\mu_A$. The arrival rates and service rates at the BS and the AP have been derived in Subsection 8.5.1.

On the other hand, the round-trip time can be expressed as

$$rtt = 2t_{wired} + t_{wireless}^{down} + t_{wireless}^{up}, \tag{8.42}$$

where t_{wired} is the transmission latency in the wired link (i.e., from the SN to the BS). $t_{wireless}^{down}$ and $t_{wireless}^{up}$ are the latencies for downlink and uplink transmissions in the integrated WWAN–WLAN link, respectively.

The term $t_{wireless}^{down}$ is given by

$$t_{wireless}^{down} = D \cdot (Q_B + \theta_S) + \delta \cdot (Q_A + \eta_S), \tag{8.43}$$

where D and δ are time slot lengths in the queueing models for the WWAN and WLAN links, respectively. θ_S and η_S are the average service times for a successful WWAN downlink transmission and for a successful WLAN downlink transmission, respectively. Q_B and Q_A are the queueing delays at the BS buffer and the AP buffer for downlink transmission, respectively. By the $M/M/1/K$ queueing model, Q_B and Q_A are given by

$$Q_B = \frac{1}{\lambda_B(1 - P_B)} \left(\frac{\rho_B}{1 - \rho_B} - \frac{\rho_B(B\rho_B{}^B + 1)}{1 - \rho_B{}^{B+1}} \right) \tag{8.44}$$

and

$$Q_A = \frac{1}{\lambda_A(1 - P_Q)} \left(\frac{\rho_A}{1 - \rho_A} - \frac{\rho_A(Q\rho_A{}^Q + 1)}{1 - \rho_A{}^{Q+1}} \right). \tag{8.45}$$

The derivations of θ_S and η_S are similar to those of $1/\mu_B$ and $1/\mu_A$, respectively, except that a successful packet transmission is assumed. θ_S is then computed as

$$\theta_S = (1 - \pi_e)$$

$$\cdot \left(\frac{1}{1 - p_{gb}p_{bb}{}^{l-1}} (p_{gg} + 2p_{gb}p_{bg} + 3p_{gb}p_{bb}p_{bg} + \cdots + l p_{gb}p_{bb}{}^{l-2}p_{bg}) \right)$$

$$+ \pi_e \left(\frac{1}{1 - p_{bb}{}^l} (p_{bg} + 2p_{bb}p_{bg} + 3p_{bb}p_{bb}p_{bg} + \cdots + l p_{bb}p_{bb}{}^{l-2}p_{bg}) \right)$$

$$= \frac{1 - \pi_e}{1 - p_{gb}p_{bb}{}^{l-1}} \left(1 + p_{gb} \left(\frac{1 - p_{bb}{}^{l-1}}{1 - p_{bb}} - l p_{bb}{}^{l-1} \right) \right) + \frac{\pi_e}{1 - p_{bb}{}^l}$$

$$\cdot \left(\frac{1 - p_{bb}{}^l}{1 - p_{bb}} - l p_{bb}{}^l \right), \tag{8.46}$$

where $1 - p_{gb}p_{bb}{}^{l-1}$ and $1 - p_{bb}{}^l$ are the probabilities that a packet transmission is successful when the WWAN link states at the last transmission of the previous packet are g and b, respectively. Also, it can be derived that

$$\eta_S = (N\lambda_M^U \eta_S (1 - \varepsilon_L^M) + 1)T_s + \frac{1}{2}(N\lambda_M^U \eta_S \overline{C_M} + \overline{C_A}|Succ)T_C + \overline{BO_A}|Succ \tag{8.47}$$

where $\overline{C_A}|Succ$ and $\overline{BO_A}|Succ$ are the average numbers of collisions and backoffs experienced by the AP when a packet is successfully transmitted over the WLAN link, respectively. They are given by

$$\overline{C_A}|Succ = 1 \cdot \frac{p_A(1 - p_A)}{1 - p_A{}^{m+1}} + 2 \cdot \frac{p_A{}^2(1 - p_A)}{1 - p_A{}^{m+1}} + \cdots + m \cdot \frac{p_A{}^m(1 - p_A)}{1 - p_A{}^{m+1}}$$

$$= \frac{1}{1 - p_A{}^{m+1}} \left(\frac{p_A(1 - p_A{}^m)}{1 - p_A} - m p_A{}^{m+1} \right) \tag{8.48}$$

and

$$\overline{BO_A}|Succ = \frac{W_0 - 1}{2} \cdot \frac{(1 - p_A{}^{m+1})}{1 - p_A{}^{m+1}} + \frac{W_1 - 1}{2} \cdot \frac{(p_A - p_A{}^{m+1})}{1 - p_A{}^{m+1}}$$

$$+ \cdots + \frac{W_m - 1}{2} \cdot \frac{(p_A{}^m - p_A{}^{m+1})}{1 - p_A{}^{m+1}}$$

$$= \sum_{i=0}^{m} \frac{W_i - 1}{2} \frac{(p_A{}^i - p_A{}^{m+1})}{1 - p_A{}^{m+1}}, \tag{8.49}$$

where $1 - p_A{}^{m+1}$ is the probability of a successful transmission by the AP.

The queueing delay for the upstream transmission is negligible. Therefore, $t_{wireless}^{up}$ is given by the transmission latency in the WWAN–WLAN link as

$$t_{wireless}^{up} = \delta \cdot \eta_S^U + \theta_S^U, \tag{8.50}$$

where θ_S^U and η_S^U are the average service times for a successful WWAN uplink transmission and for a successful WLAN uplink transmission, respectively. η_S^U can be derived as

$$\eta_S^U = ((N-1)\lambda_M^U \eta_S^U (1 - \varepsilon_L^M) + \lambda_A (1 - P_Q)\eta_S^U (1 - \varepsilon_L^A) + 1)T_s$$

$$+ \frac{1}{2}((N-1)\lambda_M^U \eta_S^U \overline{C_M} + \overline{C_M} + \lambda_A (1 - P_Q)\eta_S^U \overline{C_A})T_c + \overline{BO_M}. \qquad (8.51)$$

On the other hand, since an ideal WWAN uplink channel is assumed, θ_S^U is simply given by P_{ack}/W_{up}, where P_{ack} and W_{up} are the TFRC *ack* size and the WWAN uplink bandwidth, respectively.

Now, we can derive the TFRC throughput in steady state using the iterative algorithm given in Algorithm 5. We consider two cases: 1) there is only one TFRC flow and the sending rates of other flows are fixed, i.e., CBR flows, (referred to as *single TFRC flow case*) and 2) each MN has a TFRC flow and therefore there are N TFRC flows in a mobile hotspot (referred to as the *multiple TFRC flows case*). For the single TFRC flow case, let λ_M' be the downlink transmission rate of the tagged MN with a TFRC flow. With the sending rate λ_{Init} initialized to 1.0, λ_T can be determined as $\lambda_T = \lambda_{Init} + (N-1)\lambda_F$, where λ_F is the constant sending rate (in packets/s) of an MN except for the tagged MN with a TFRC flow. In the sequel, *rtt* and *p* are computed using λ_T (8.40) and (8.42). T_0 is set to 4*rtt* [143] and a new TFRC sending rate λ_M' is calculated in line 5 of Algorithm 5. In lines 6–12, using a sufficiently small value ϵ, λ_M' is repeatedly calculated until it converges. On the other hand, for the multiple TFRC flows case, λ_T is set according to $\lambda_T \leftarrow N\lambda_{Init}$ in lines 2 and 8, and λ_M' is repeatedly computed using (7.1). Consequently, the throughput of a TFRC flow, T, can be computed as $T = \lambda^*(1 - p^*)$ [168], where λ^* and p^* are the TFRC sending rate and the packet loss rate in steady state, respectively.

8.5.3   Numerical results

In the simulation, the network topology is the same as that shown in Figure 8.15. The following parameters are used unless otherwise explicitly stated. The carrier frequency (f_c) of the WWAN link is 900 MHz. The payload sizes of a data packet and an *ack* packet are fixed to 250 bytes and 50 bytes, respectively. The default downlink and uplink WWAN bandwidths are 400 Kbps and 80 Kbps, respectively. Hence, the transmission time of a packet (i.e., data or *ack* packet) over the WWAN link is 5 ms. The default velocity (v) and fading margin (F) are 20 m/s and 10 dB, respectively. The default values of retransmission limits (i.e., l and m) for the WWAN and WLAN links are 3 and 5, respectively. The number of MNs within a vehicle is varied from 1 to 40 and its default value is 20. t_{wired} is set to 20 ms. The default system sizes of the AP buffer (Q) and the BS buffer (B) are 10 (packets). The parameters for WLAN follow those of the IEEE 802.11b specification in [167] and the data rate is 11 Mbps. To validate analytical results, simulations are performed using the *NS-2* simulator.

Algorithm 5 Algorithm to derive TFRC throughput [164]. Reproduced by permission of ©2008 IEEE.

1: $\lambda_{Init} \leftarrow 1$;
2: $\lambda_T \leftarrow \lambda_{Init} + (N-1)\lambda_F$;
3: Calculate rtt and p using λ_T, (8.40), and (8.42);
4: $T_0 \leftarrow 4rtt$;
5: Calculate λ'_M using (7.1);
6: **while** $|\lambda'_M - \lambda_{Init}| \geq \epsilon$ **do**
7:    $\lambda_{Init} = (\lambda'_M + \lambda_{Init})/2$;
8:    $\lambda_T \leftarrow \lambda_{Init} + (N-1)\lambda_F$;
9:    Calculate rtt and p using λ_T, (8.40), and (8.42);
10:    $T_0 \leftarrow 4rtt$;
11:    Calculate λ'_M using (7.1);
12: **end while**

Effects of v and N

Figure 8.16 shows the effect of velocity (v) on the TFRC throughput T for the multiple TFRC flows case. Note that the impact of v on the physical layer, e.g., synchronization error, is not considered. It can be seen that, given the same level of fading margin, T increases as v increases. This observation is consistent with that in Chapter 7.

Figure 8.16 also shows that T decreases as the number of MNs sharing the WLAN increases. This is because a large N leads to a higher packet loss rate due to more channel collisions in the WLAN and a longer queueing delay. However, the effect of N on T is not significant, especially when v is low. This is because each TFRC flow adapts to the network conditions. In other words, when N is large, the TFRC flows will reduce their sending rates if higher packet loss rates are observed. In short, the flow and congestion control mechanisms of TFRC can intelligently adjust the sending rate to effectively mitigate network congestion, and efficiently utilize network resources independent of N.

The effects of v and N for the single TFRC flow case are illustrated in Figure 8.17. We consider two values for λ_F: 4 packets/s (i.e., 8 Kbps) and 8 packets/s (i.e., 16 Kbps). Similar to Figure 8.16, the TFRC throughput increases with the increase of v, and the increasing rate becomes stable when v exceeds a certain point. On the other hand, the effect of N is clear in the single TFRC case, especially when λ_F is high. From both Figures 8.16 and 8.17, the analytical results match the simulation ones.

The TFRC throughputs in the single and multiple TFRC cases are compared in Figure 8.18. For the multiple TFRC case, the TFRC throughput is not highly sensitive to N due to TFRC flow and congestion control mechanisms. However, in the single TFRC case, the TFRC throughput drastically decreases as N increases. Especially,

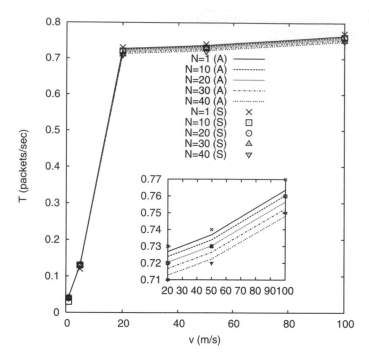

Figure 8.16 T vs. v: multiple TFRC flows (A, analytical; S, simulation) [164]. Reproduced by permission of ©2008 IEEE.

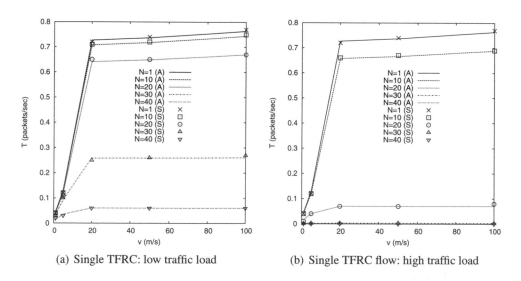

(a) Single TFRC: low traffic load   (b) Single TFRC flow: high traffic load

Figure 8.17 T vs. v: single TFRC flow (A, analytical; S, simulation) [164]. Reproduced by permission of ©2008 IEEE.

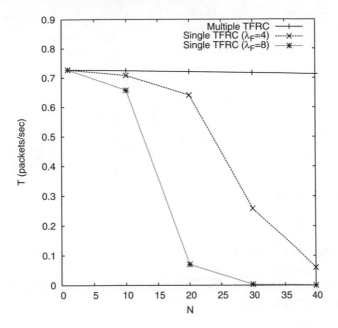

Figure 8.18   Effect of N: multiple TFRC flows vs. single TFRC flow [164].
Reproduced by permission of ©2008 IEEE.

for $\lambda_F = 8$ (packets/s), the TFRC throughput reduces below 0.001 (packets/s) when N is larger than 30, which indicates that a TFRC flow cannot be effectively supported. This is because the high traffic load incurs more packet losses (due to buffer overflow and channel collisions in WLAN) and a longer queueing delay, and thus degrades the TFRC throughput. Therefore, it is conjectured that an admission control algorithm to limit the number of MNs in a mobile hotspot should be deployed to provide satisfactory QoS.

Effects of l and m

Retransmissions up to $l - 1$ and m times are deployed in the WWAN and WLAN links, respectively, and they affect the packet loss probability and the round-trip time. Figure 8.19 shows the effects of l and m on the TFRC throughput. As shown in Figure 8.19(a), when l increases, a higher TFRC throughput can be obtained, since the packet loss rate can be significantly reduced for a large l. However, a larger l may not be preferable for delay-sensitive multimedia applications because it will increase the end-to-end delay. Therefore, an optimal l should be determined by considering the tradeoff between latency and throughput. On the other hand, the effect of m is not significant, as shown in Figure 8.19(b). Since the WWAN is most likely to be

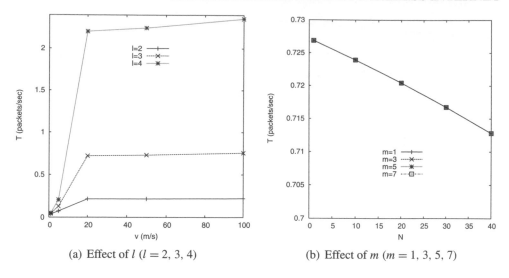

(a) Effect of l ($l = 2, 3, 4$)   (b) Effect of m ($m = 1, 3, 5, 7$)

Figure 8.19 Effect of retransmission limits: multiple TFRC flows [164]. Reproduced by permission of ©2008 IEEE.

Table 8.5   TFRC throughput for different (B, Q): multiple TFRC flows [164]. Reproduced by permission of ©2008 IEEE.

N	(5, 10)	(10, 10)	(20, 10)	(10, 5)	(10, 20)
1	0.723863731	0.726856708	0.726856708	0.726856708	0.726856708
10	0.72387965	0.723915679	0.723915679	0.723915679	0.723915679
20	0.719395134	0.720448182	0.720448182	0.720448182	0.720448182
30	0.7097508	0.716753914	0.716753914	0.716753914	0.716753914
40	0.689423247	0.712811896	0.712813894	0.71281089	0.712811896

the bottleneck and TFRC flows can adjust the sending rate to mitigate congestion in WLAN, packet losses due to collisions are rare for the multiple TFRC case.

Effects of B and Q

The BS and AP buffer sizes determine the queueing delay and the packet loss rate. Table 8.5 shows the TFRC throughput when different B and Q are employed. It can be seen that the TFRC throughput remains at almost the same value when Q varies and B is fixed. That is, the effect of Q on the TFRC throughput is quite limited. As mentioned before, for the multiple TFRC flows case, network congestion in the WLAN is not significant due to TFRC's flow and congestion control mechanisms. Also, the WLAN bandwidth is much larger than the WWAN bandwidth. Therefore, the AP is unsaturated, i.e., $\rho_A < 1.0$. Consequently, under a lightly loaded WLAN

Table 8.6 TFRC throughput for different (B, Q): single TFRC flow $(\lambda_F = 4)$ [164]. Reproduced by permission of ©2008 IEEE.

N	(5, 10)	(10, 10)	(20, 10)	(10, 5)	(10, 20)
1	0.726856708	0.726856708	0.726856708	0.726856708	0.726856708
10	0.627093524	0.70781588	0.70784891	0.707812876	0.70781588
20	0.214665207	0.640683244	0.674254834	0.640594151	0.640683244
30	0.06815438	0.257965069	0.594110782	0.257902797	0.257965069
40	0.021640276	0.060273945	0.150578214	0.060263745	0.060273945

Table 8.7 TFRC throughput for different (B, Q): single TFRC flow $(\lambda_F = 8)$ [164]. Reproduced by permission of ©2008 IEEE.

N	(5, 10)	(10, 10)	(20, 10)	(10, 5)	(10, 20)
1	0.726856708	0.726856708	0.726856708	0.726856708	0.726856708
10	0.242244246	0.657371106	0.678541517	0.657298033	0.657371106
20	0.024378619	0.069964461	0.188789698	0.069952532	0.069964461
30	0.002428256	0.003178217	0.002375052	0.003177681	0.003178217
40	0.000521237	0.000435947	0.000278681	0.0004359	0.000435947

condition, the variation of Q has no significant impact on the TFRC throughput. On the other hand, throughput degradation can be observed when B is reduced from 20 (or 10) to 5. It is noted that the degradation is more pronounced when N is large.

Tables 8.6 and 8.7 show the effects of B and Q for the single TFRC flow case. It can be seen that the effect of Q is not significant regardless of λ_F. For $\lambda_F = 4$ (packets/s), significant throughput degradation is observed for a small value of B. This is because a small BS buffer induces more packet losses due to buffer overflow. On the other hand, for $\lambda_F = 8$ (packets/s), a large B gives a higher TFRC throughput when N is less than 20. However, when N is equal to or larger than 20, a small buffer can increase the TFRC throughput. Obviously, a large buffer can reduce the packet loss rate while it increases the round-trip time due to the queueing delay. Since the WWAN link is not congested when the number of CBR flows is small, the increase in the round-trip time due to a large buffer is not significant and therefore a large buffer is better to improve the TFRC throughput. On the contrary, when there are many CBR flows (i.e., N is large), increasing the BS buffer size cannot significantly reduce the packet loss rate, while the queueing delay will be substantially prolonged. Therefore, it is not desirable to use a larger BS buffer. From these observations, it can be concluded that the dimensioning of the BS buffer size is critical to improve the throughput of the TFRC flow in mobile hotspots. In addition, it can be shown that the load of non-responsive traffic has a significant impact on the TFRC throughput in mobile hotspots.

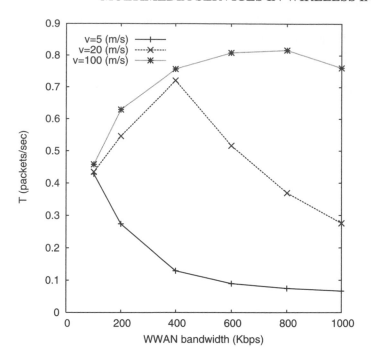

Figure 8.20 Effect of WWAN bandwidth: multiple TFRC flows [164]. Reproduced by permission of ©2008 IEEE.

Effect of WWAN/WLAN Bandwidth

Figure 8.20 shows the effect of the WWAN downlink bandwidth with different velocities. For $v = 5$ m/s, the TFRC throughput decreases as the allocated WWAN bandwidth increases. This counter-intuitive observation can be explained as follows. In the WWAN link, a time slot equals a packet transmission time over the WWAN link. Therefore, with higher WWAN bandwidth, the transmission time of a packet is smaller and thus the time slot duration is short. When v is low, the channel coherence time in the WWAN link is long and a shorter time slot under the long channel coherence time leads to a longer burst of transmission errors, which may not be recovered by the truncated ARQ scheme. Thus, more packet losses are observed by the transport layer. Consequently, at a low velocity, even though the round-trip time can be shortened by a large WWAN bandwidth, the TFRC throughput is degraded due to the high packet loss rate. This undesirable situation can be improved by exploiting diversity, e.g., retransmitting the corrupted packets at another band/subcarrier in a FDMA/OFDM system, using delayed retransmission in a TDMA system, etc., which are beyond the scope of this book.

On the other hand, for $v = 20$ m/s and $v = 100$ m/s, it can be observed that the TFRC throughput can be improved by allocating more bandwidth, since the channel

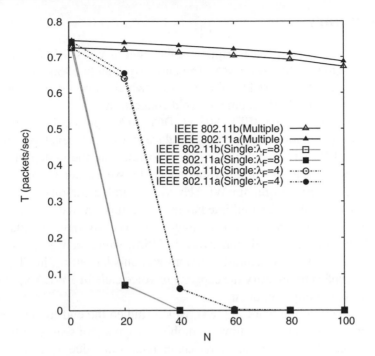

Figure 8.21 Effect of WLAN bandwidth: multiple TFRC flows [164]. Reproduced by permission of ©2008 IEEE.

coherence time is not too long. Another interesting result is that there is an optimal WWAN bandwidth to maximize the TFRC throughput. Therefore, when the allocated bandwidth is larger than the optimal value, the TFRC throughput is reduced due to the burstiness of transmission errors. In addition, the optimal WWAN bandwidth increases with v, i.e., given the wireless channel profile, it is preferable to allocate more WWAN bandwidth to maximize the TFRC throughput according to v.

To investigate the effect of the WLAN bandwidth, we consider an IEEE 802.11a [169] WLAN supporting a high data rate of 54 Mbps. As shown in Figure 8.21, the TFRC throughput can be improved when IEEE 802.11a is used. However, the improvement is not significant even though the bandwidth of IEEE 802.11a is much larger than that of IEEE 802.11b. Since the bottleneck of mobile hotspots is the WWAN link in general, the WLAN load is not sufficiently heavy to require more bandwidth. Only if larger bandwidth is allocated in the WWAN such that the bottleneck is shifted to WLAN, the gain of higher data rate WLANs with IEEE 802.11a/g will be significant. From Figure 8.21, it can be seen that limiting the number of unresponsive flows is more important than increasing the WLAN bandwidth for improving the TFRC throughput; this result demonstrates the necessity of admission control in mobile hotspots.

8.6   Summary

In this chapter, we start with a description of the random access MAC protocols, from pure Aloha, slotted Aloha, to CSMA. The currently widely adopted MAC protocol for WLANs is the IEEE 802.11 DCF protocol, which is based on CSMA and the random exponential backoff scheme to avoid congestion.

The performance of the IEEE 802.11 DCF MAC protocol has been heavily investigated in the literature. First, we introduce the model developed by Bianchi: a bi-dimensional DTMC model which is used to calculate the system throughput as a function of the number of saturated stations [151]. We further extend the results to obtain the average service time for WLANs with saturated stations.

Second, since senders for multimedia traffic are unsaturated in general, and the stations (AP and MNs) in an infrastructure-based WLAN have unbalanced traffic load, we discuss the analytical framework in [152], which can be used to quantify system performance with unsaturated stations and unbalanced traffic. The analytical framework is applied to quantify the capacity of voice calls in a WLAN, as VoWLAN becomes ever increasingly popular.

Finally, we present the work in [164], which studied the end-to-end performance of TFRC-controlled multimedia flows in mobile hotspots (consisting of both WWAN and WLAN links). This is a particularly challenging topic due to the interactions of the end-to-end congestion control protocol and the service time variations resulting from transmission errors in the WWAN and the contentions in the WLAN. The analytical framework developed in [164] is based on discrete-time queueing models. Given the queueing models, the TFRC throughput in a mobile hotspot can be derived. Analytical and simulation results demonstrate that the TFRC throughput is largely affected by the WWAN channel condition and bandwidth. In addition, since the WWAN–WLAN link is shared by multiple MNs, the traffic load has a significant impact on the TFRC throughput in mobile hotspots. Therefore, a suitable bandwidth allocation method and an admission control algorithm are necessary to support multimedia applications with QoS guarantee in mobile hotspots. The analytical and simulation results presented can be used as a guideline for effective admission control in mobile hotspots, which can be an interesting future research topic.

8.7   Problems

1. Consider a non-persistent *slotted CSMA* protocol: time is slotted, and the duration of a slot T equals the transmission time of a frame. For a frame to be transmitted, if the channel is sensed idle, it can be transmitted at the beginning of the next slot; if the channel is sensed busy, it randomly backs off.

   (a) Would the non-persistent slotted CSMA outperform the non-persistent CSMA in a local area network? Explain why.

    (b) How would you set the protocol parameter T to improve the efficiency of the non-persistent slotted CSMA protocol?

2. There are N MNs in a single-hop IEEE 802.11b WLAN. Channel data rate is 11 Mbps, and all other parameters follow the standard. All nodes are saturated, i.e., they always have packets to send. Assume that, for each transmission, there is a probability of a that the transmission fails due to channel impairments. The DCF MAC protocol and the basic access mode are used. Extend Bianchi's model to answer the following questions:

    (a) Would the collision probability in the WLAN be affected by a, the probability of failed transmission due to channel impairments? Derive the collision probability in the WLAN.

    (b) Derive the throughput in the WLAN.

    (c) Derive the average service time for packet transmissions in the WLAN.

3. The following strategy has been proposed to provide differentiated services for different classes of traffic in the WLAN with DCF MAC protocol: the minimum contention window size (CW_{min}) for higher priority traffic is smaller than that for lower priority traffic. Consider a single-hop IEEE 802.11 WLAN that deploys the above strategy and supports two classes of traffic. For class I traffic, the CW_{min} is 16; for class II traffic, the CW_{min} is 32. There are eight saturated stations in the WLAN. Half of them contend the channel for class I traffic, and the remaining for class II traffic.

    (a) Derive the collision probabilities for class I and class II packet transmissions, respectively.

    (b) Derive the service times for class I and class II packets, respectively.

    (c) Derive the throughputs for stations with class I traffic and class II traffic, respectively.

4. The number of MN in an area of size A follows the Poisson distribution with the average number equal to ρA. Assume that both the transmission range and the sensing range are equal to R. Assume $\rho \pi R^2 >> 1$ so that there are no isolated nodes. RTS/CTS are used for packet transmission. The transmission time of the data packet and that of RTS and CTS are t_D, t_R and t_C, respectively. The p-persistent CSMA scheme is used: if the channel is sensed idle for *DIFS*, for every time slot with duration T, each node will attempt to transmit an RTS with probability p to one of its neighbors; if the RTS and CTS messages are exchanged successfully, the node will transmit a packet followed by an ACK. Wireless channel condition is assumed ideal such that all transmitted frames can be received error-free if there is no collision.

    (a) What is the vulnerable period of a packet transmission?

    (b) What is the average number of hidden terminals?

    (c) What is the collision probability?

    (d) What is the average number of transmissions for each packet until success?

5. Consider an infrastructure-based WLAN which deploys IEEE 802.11b DCF MAC with the basic access mode. There are one AP and N MNs. The AP and the CNs are connected through wired links with sufficient bandwidth. Multimedia connections are established between MNs and CNs through the AP. The traffic from the CNs to the MNs is CBR traffic with arrival rate of a packets per second; and the reverse direction traffic is CBR traffic with arrival rate of $0.2a$ packets per second.

    (a) Derive the collision probability for packet transmissions at the AP.

    (b) Derive the collision probability for packet transmissions at the MNs.

    (c) Derive the maximum value of a that the WLAN can support without excessive delay and loss.

6. Consider a mobile hotspot as described in Subsection 8.5.3. Let N_1 TCP connections and N_2 TFRC connections be established between $N_1 + N_2$ MNs and $N_1 + N_2$ CNs through the AP in a WLAN, and let the AP be connected to the BS through a WWAN link. Assume that the size of $rwnd$s of the TCP flows are always larger than their $cwnd$s, and all flows have the same values of round-trip times (without queueing delay).

    (a) Derive the average queueing delay at the AP.

    (b) Derive the throughputs of the TFRC flows.

    (c) Derive the throughputs of the TCP flows.

    (d) How would you allocate the WWAN link bandwidth to fully utilize the WWAN link and maximize the total throughput in the WLAN?

Appendices

Appendix A

TCP and AQM Overview

A.1 TCP Protocol

The TCP protocol provides connection-oriented, end-to-end reliable, stream-like service over the connectionless, unreliable, best-effort IP protocol.

A.1.1 TCP connection management

Before the two end-systems begin to exchange information, a TCP connection is established first to associate them. As shown in Figure A.1, three-way handshakes between the client and the server open a bi-directional data channel: the client initiates the TCP connection by sending out a synchronization (SYN) segment; the server responds with a SYN segment containing an acknowledgment (*ack*) of the client's SYN segment; and the client sends an *ack* of the server's SYN segment. TCP data starts to transfer thereafter.

A graceful close of the connection requires two-way handshakes that close the data channel of each direction individually: when one end has delivered all its data and has no more data to send, a finish (FIN) segment is constructed to notify the other end, followed by the acknowledgment to the FIN segment, and vice versa.

A.1.2 TCP error control

The IP protocol provides best-effort packet transmission in which packets may be lost, corrupted, duplicated and reordered during transmission. The end-to-end data integrity depends on the error control function of the TCP protocol via the use of sequence numbers and acknowledgments.

A TCP sender packages data into segments, and each segment is identified by a sequence number. Upon receiving a segment successfully, the receiver sends an *ack* that contains the value of the next sequence number the receiver is expecting to receive, as shown in Figure A.2. When the TCP sender transmits a segment containing data, it puts a copy of the segment in a retransmission queue and starts

Multimedia Services in Wireless Internet: Modeling and Analysis   Lin Cai, Xuemin Shen and Jon W. Mark
© 2009 John Wiley & Sons, Ltd

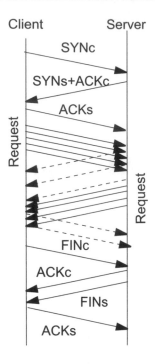

Figure A.1 TCP connection management.

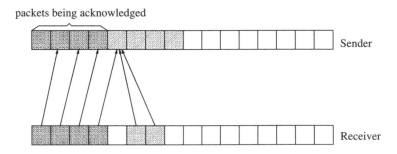

Figure A.2 Acknowledgment scheme.

a timer; when the *ack* for that data is received, the segment is deleted from the queue. If the *ack* is not received before the timer runs out (a timeout event), the segment is retransmitted.

To provide basic protection against transmission errors, TCP includes a 16-bit checksum field in its header. The receiver recalculates the checksum to identify corrupted segments that will be discarded by the receiver. The sender then retransmits the corrupted segments when timeout occurs.

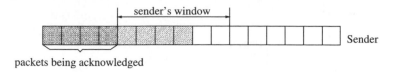

Figure A.3 Sliding window mechanism.

Duplicated segments can be identified by the TCP receiver according to the sequence number; the receiver simply discards the duplicated ones and no further recovery is needed at the sender side.

For reordered segments, the receiver can recover the correct order in its buffer, using the sequence numbers.

A.1.3   TCP flow control and congestion control

In the Internet, a TCP connection can be established between any two end-systems with different computational power, memory, etc. To avoid a fast sender overrunning a slow receiver, a sliding-window-based flow control scheme is used. As shown in Figure A.3, the window is defined in the sequence space. The TCP sender can only transmit a window size of segments without receiving the acknowledgment from the receiver. The lower edge of the window is the sequence number of the first unacknowledged segment, so the window slides once a new *ack* is received. The receiver informs the sender of the maximum window size, so the TCP sending rate is bounded by the receiver advertised window (*rwnd*) size over the round-trip time.

To coordinate the end-systems and the network, the window-based congestion control algorithm was incorporated into the TCP protocol in the late 1980s when a series of congestion collapses were observed, even at that time when the Internet was relatively small [170]. The main principle of window-based congestion control is *packet conservation* (also known as *acknowledgment self-clocking*): a new segment is injected into the network only after an old one has left. This is an ideal case in the network and session equilibrium state. However, changes in the available network resources and the number of sessions always deviate TCP away from the ideal case. Therefore, TCP's dynamic congestion control is necessary to maintain network stability [85], and the Internet evolution in the last two decades has proved the effectiveness of the TCP congestion control mechanism.

A TCP sender uses a congestion window (*cwnd*) to probe for available bandwidth and respond to network transient congestions. The TCP sender's window size is the minimum of *rwnd*, *cwnd* and the sender buffer size. The TCP sender probes for available network resources with a small *cwnd* initially, after a timeout or being idle for a while. In order to reach the target equilibrium quickly, TCP increases *cwnd* exponentially every *rtt* until *cwnd* is above the *slow start threshold* (*ssthresh*) or when

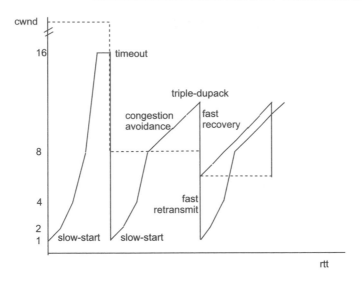

Figure A.4 TCP congestion control.

congestion occurs. This is known as *slow start*, as shown in Figure A.4. The initial value of *cwnd* is one MSS, and *ssthresh* is set to *rwnd* initially.

When *cwnd* is larger than *ssthresh*, TCP probes for the unused bandwidth conservatively, i.e., increasing *cwnd* by one segment per *rtt* (without the delayed acknowledgment policy), which is referred to as *congestion avoidance*, until eventually congestion occurs. For the mainstream TCP variants, timeout and duplicated acknowledgment (*dupack*) are two common congestion indicators. Timeout indicates severe congestion, and forces TCP to reinitialize *cwnd* and halve *ssthresh*. Normally, the occurrence of three *dupacks* is considered to signify moderate congestion, so *cwnd* is halved (or more precisely, reduced to a half of the current outstanding data) and *ssthresh* is set to *cwnd*; this is referred to as *exponential backoff*.

With the wide adoption of TCP/IP protocols in the Internet, TCP congestion control also continues its evolution. New algorithms, e.g., Partial Acknowledgment and Selective Acknowledgment, have been proposed and incorporated into TCP. With the limited information available from cumulative acknowledgments, a TCP sender can learn about only a single lost packet per round-trip time. A SACK mechanism, combined with a selective repeat retransmission policy, can help to overcome these limitations. The TCP receiver sends back SACK packets to inform the sender that the sent data has been received. The sender can then retransmit the missing data segments to efficiently recover multiple packet losses in one window. The interested reader is referred to [171, 172] for additional information.

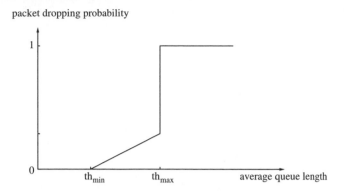

Figure A.5 RED queue packet dropping probability.

A.2   Active Queue Management

As packet losses are often due to network congestion, end-systems infer network congestion conditions when packet loss occurs. For Internet intermediate systems, *Drop-Tail* routers drop incoming packets whenever their output buffer overflows. Drop-Tail queueing is known to produce bursty packet losses and a bias against flows with long *rtt* and small packets.

Modern active queue management (AQM)-capable routers can signify congestion even before buffer overflow actually occurs [173]. RED is a well-known AQM scheme and it has been deployed by major router vendors like Cisco and Juniper [7]. RED-capable routers monitor the moving-average of queue length (and the speed at which it increases). If the average queue length is below a lower threshold, no packet is dropped; if it is above an upper threshold, all packets are dropped. When the average queue length is between the two thresholds, packets are dropped with a certain probability, which is a function of the queue length, as shown in Figure A.5. With RED-capable routers, the packet loss probability of a flow tends to be proportional to its packet sending rate, and the congestion indicators can be distributed to coexisting flows more fairly.

Furthermore, the ECN scheme [120] has been proposed in both the network and transport layers. ECN-capable routers can advertise the incoming congestion without dropping packets, and ECN-capable end-systems can exercise congestion control proactively.

Appendix B

Datagram Congestion Control Protocol Overview

Since TCP meets great challenges in supporting multimedia applications, the Datagram Congestion Control Protocol (DCCP) has been proposed to regulate and support multimedia applications [174]. If DCCP is to be preferred over the unresponsive UDP protocol, its overhead should be small. Thus, DCCP provides only a minimal functionality of congestion control. For different applications, there are different types of congestion control mechanisms, such as TCP-like congestion control (DCCP-2) and TFRC congestion control (DCCP-3), to choose.

B.1   DCCP-2: TCP-like Congestion Control

Reference [83] introduces the basic features of DCCP-2 as follows:

1. The sender uses a *cwnd* to regulate the number of DCCP data packets being sent without acknowledgment. The sender also sets an Acknowledgment Ratio (AckRatio) option specifying how many data packets are to be covered by a DCCP *ack* from the receiver. The default AckRatio is two.

2. For every AckRatio data packet transmitted by the sender, the receiver sends a DCCP *ack*, which uses a sequence number and contains an Acknowledgment Vector (AckVector). Unlike the TCP *ack*, the DCCP *ack* does not have a cumulative acknowledgment field since DCCP does not guarantee all packets are transmitted successfully and in order.

3. Upon receiving a DCCP *ack*, the sender examines the AckVector and learns about packet losses. It increases or decreases the *cwnd* accordingly; this is similar to SACK-based TCP (TCP-SACK) congestion control. But the DCCP

sender does not retransmit unacknowledged packets for unreliable datagram delivery services.

4. Because DCCP *ack*s use sequence numbers, the sender has direct information about the fraction of lost or marked *ack*s. The sender responds to lost or marked *ack*s by modifying the AckRatio sent to the receiver. In this way, the sending rate of *ack* flow can be regulated to be roughly TCP-friendly.

5. The sender acknowledges the receiver's acknowledgments at least once per congestion window. The sender sends a DCCP-DataAck packet that includes an Acknowledgment Number in the header. (The *ack*s-of-*ack*s is a new feature in DCCP compared with TCP. However, as reported in [175], it is too complex to implement in practice.)

6. The sender estimates round-trip times and calculates a timeout value. The timeout is used to determine if a new DCCP-Data packet can be transmitted when the sender has been limited by the *cwnd* and no feedback has been received from the receiver.

7. Each DCCP-data packet is sent as ECN-capable.

B.2 DCCP-3: TFRC Congestion Control

DCCP-3 has been designed for the applications that want a TCP-friendly sending rate, while minimizing abrupt rate changes. The basic procedures of DCCP-3 are given in [176]:

1. The sender transmits data segments, where the sending rate is governed by the allowed transmit rate as the TFRC mechanism specified in IETF RFC 3448 [145].

2. The receiver sends *ack*s at least once per round-trip time acknowledging the data segments, unless the sender is sending at a rate of less than one packet per round-trip time.

3. Upon receiving *ack*s, the sender updates its allowed transmit rate and estimates round-trip times according to the TFRC mechanism specified in RFC 3448.

References

[1] Keshav S. Why cell phones will dominate the future Internet. *ACM Computer Communication Review*, 35(2):83–86, 2005.

[2] Jaffe J.M. Bottleneck flow control. *IEEE Transactions on Communications*, 29(7):954–962, 1981.

[3] Tanenbaum A.S. *Computer Networks*. Prentice Hall PTR, 4th edn, 2002. ISBN: 0130661023.

[4] Kurose J.F. and Ross K.W. *Computer Networking: A Top-Down Approach Featuring the Internet*. Addison-Wesley, 2004. ISBN: 0321227352.

[5] Leon-Garcia A. and Widjaja I. *Communication Networks: Fundamental Concepts and Key Architectures*. McGraw-Hill Companies, 2003. ISBN: 007246352X.

[6] Floyd S. and Fall K. Promoting the use of end-to-end congestion control in the Internet. *IEEE/ACM Transactions on Networking*, 7(4):458–472, 1999.

[7] Floyd S. and Jacobson V. Random Early Detection gateways for congestion avoidance. *IEEE/ACM Transactions on Networking*, 1(4):397–413, 1993.

[8] Weighted random early detection. Cisco documentation. http://www.cisco.com/en/US/docs/ios/12_0s/feature/guide/fswfq26.html.

[9] Pan R., Prabhakar B. and Psounis K. CHOKe, a stateless active queue management scheme for approximating fair bandwidth allocation. In *Proceedings of IEEE Infocom'00*, March 2000.

[10] Stewart R., Xie Q., Morneault K., *et al.* Stream control transmission protocol. *IETF RFC 2960*, Oct. 2000.

[11] Schulzrinne H., Casner S., Frederick R. and Jacobson V. RTP: A transport protocol for real-time applications. *IETF RFC 1889*, Jan. 1996.

[12] Floyd S., Handley M. and Kohler E. Problem statement for DCCP (Aug. 2005). Available at www.icir.org/kohler/dcp/draft-ietf-dccp-problem-03.txt.

[13] Goyal V.K. Multiple description coding: compression meets the network. *IEEE Signal Processing Magazine*, 18:74–93, 2001.

[14] Li W. Overview of fine granularity scalability in MPEG-4 standard. *IEEE Transactions on Circuits and Systems for Video Technology*, 11(3):301–317, 2001.

[15] Radha H.M., van der Schaar M. and Chen Y. The MPEG-4 fine-grained scalable video coding method for multimedia streaming over IP. *IEEE Transactions on Multimedia*, 3(1):53–68, 2001.

[16] Wu D., Hou Y.T., Zhu W., Zhang Y. and Peha J. Streaming video over the Internet: approaches and directions. *IEEE Transactions on Circuits and Systems for Video Technology*, 11(3):282–300, 2001.

[17] Kondi L.P. and Katsaggelos A.K. Joint source-channel coding for scalable video using models of rate-distortion functions. In *Proceedings of the 2001 IEEE International Conference on Acoustics, Speech, and Signal Processing*, vol. 3, pp. 1377–1380, 2001.

[18] Puri R., Lee K.W., Ramchandran K. and Bharghavan V. An integrated source transcoding and congestion control paradigm for video streaming in the Internet. *IEEE Transactions on Multimedia*, 3(1):18–32, 2001.

[19] Ayanoglu E., Paul S., LaPorta T.F., *et al.* AIRMAIL: A link-layer protocol for wireless networks. *Wireless Networks*, 1(1):47–69, 1995.

[20] Wong J.W.K. and Leung V.C.M. Improving end-to-end performance of TCP using link-layer retransmissions over mobile internetworks. In *Proceedings of IEEE ICC'99*, pp. 324–328, 1999.

[21] Parsa C. and Garcia-Luna-Aceves J.J. Improving TCP performance over wireless network at the link layer. *Mobile Networks & Applications*, 5(1):57–71, 2000.

[22] Jakes W.C. *Microwave Mobile Communications*. John Wiley & Sons, New York, 1974.

[23] Foerster J., *et al.* Channel modeling subcommittee report (final). IEEE P802.15.3a Working Group, P802.15-03/02490r1-SG3a, Feb. 2003.

[24] Gilbert E.N. Capacity of a burst-noise channel. *Bell Systems Technical Journal*, 39:1253–1265, 1960.

[25] Elliot E.O. Estimates of error rates for codes on burst-noise channels. *Bell Systems Technical Journal*, 42:1977–1997, 1963.

[26] Wang H.S. and Moayeri N. Finite-state Markov channel – a useful model for radio communication channels. *IEEE Transactions on Vehicular Technology*, 44(1):163–171, 1995.

[27] Zhang Q. and Kassam S.A. Finite-state Markov model for Rayleigh fading channels. *IEEE Transactions on Communications*, 47(11):1688–1692, 1999.

[28] Mark J.W. and Zhuang W. *Wireless Communications and Networking*. Prentice Hall, 2003.

[29] Rosenthal J.S. Convergence rates of Markov chains. *SIAM Review*, 37(3):387–405, 1995.

[30] Zhang R. and Cai L. Packet-level channel model for wireless OFDM systems. In *IEEE Globecom'07*, pp. 5215–5219, Washington, DC, USA, Nov./Dec. 2007.

[31] Kang Z., Yao K. and Lorenzelli F. Nakagami-*m* fading modeling in the frequency domain for OFDM system analysis. *IEEE Communication Letters*, 7(10):484–486, 2003.

[32] Yacoub M.D., Vargas J.E.B. and de R. Guedes L.G. On higher order statistics of the Nakagami-*m* distribution. *IEEE Transactions on Vehicular Technology*, 48(3):790–794, 1999.

[33] Iskander C. and Mathiopoulos P.T. Fast simulation of diversity Nakagami fading channels using finite-state Markov models. *IEEE Transactions on Broadcasting*, 49(3):269–277, 2003.

[34] Wang H.S. and Chang P.-C. On verifying the first-order Markovian assumption for a Rayleigh fading channel. *IEEE Transactions on Vehicular Technology*, 45(2):353–357, 1996.

[35] Zhuang W., Shen X. and Bi Q. Ultra-wideband wireless communications. *Wireless Communications and Mobile Computing*, 3(6):663–685, 2003.

[36] Mathur R., Klepal M., McGibney A. and Pesch D. Influence of people shadowing on bit error rate of IEEE802.11 2.4GHz channel. In *Proceedings of the 1st IEEE International Symposium on Wireless Communication Systems*, Mauritius, Sept. 2004.

[37] Klepal M., Mathur R., McGibney A. and Pesch D. Influence of people shadowing on optimal deployment of WLAN access points. In *Proceedings of the 60th IEEE Vehicular Technology Conference (VTC)*, Los Angeles, CA, USA, Sept. 2004.

[38] Molisch A. Time variance for UWB wireless channels. IEEE P802.15-02/461r0-SG3a, Nov. 2002.

[39] Schell S.V. Analysis of time variance of a UWB propagation channel. IEEE P802.15-02/452r0-SG3a, Nov. 2002.

[40] Zhang R. and Cai L. Modeling UWB indoor channel with shadowing processes. (2007) In *Proceedings of QShine'07*, Vancouver, BC, Canada, Aug. 2007.

[41] Spence Q., Jeffs B., Jensen M. and Swindlehurst A. Modeling the statistical time and angle of arrival characteristics of an indoor multipath channel. *IEEE Journal on Selected Areas in Communication*, 18(3):347–360, 2000.

[42] Cramer R., Scholtz R. and Win M.   Evaluation of an ultra-wide-band propagation channel. *IEEE Transactions on Antennas and Propagation*, 50(5):561–570, 2002.

[43] Zhang R. and Cai L. A packet-level model for UWB channel with people shadowing process based on angular spectrum analysis. *IEEE Transactions on Wireless Communications*, 2009. Under minor revision.

[44] Saleh A. and Valenzuela R.   A statistical model for indoor multipath propagation. *IEEE Journal on Selected Areas in Communication*, 5(2):128–137, 1987.

[45] IEEE P802.15.3a Working Group. Multi-band OFDM physical layer proposal for IEEE 802.15 task group 3a. IEEE P802.15-03/268r3, Mar. 2004.

[46] Nguyen G., Noble B., Katz R. and Satyanarayanan M. A trace-based approach for modeling wireless channel behavior. In *Proceedings of the 1996 Winter Simulation Conference*, pp. 597–604, Coronado, CA, USA, 1996.

[47] Camp T., Boleng J. and Davies V. A survey of mobility models for ad hoc network research.   *Wireless Communications & Mobile Computing (WCMC): Special issue on Mobile Ad Hoc Networking: Research, Trends and Applications*, 2(5):483–502, 2002.

[48] Yoon J., Liu M. and Noble B. Random waypoint considered harmful. In *Proceedings of IEEE Infocom'03*, vol. 2, pp. 1312–1321, San Francisco, CA, USA, April 2003.

[49] Lin G., Noubir G. and Rajamaran R. Mobility models for ad-hoc network simulation. In *Proceedings of IEEE Infocom'04*, pp. 454–463, Hong Kong, China, Mar. 2004.

[50] Guan Y.L. and Turner L.F. Generalised FSMC model for radio channels with correlated fading. *IEE Proceedings – Communications*, 146(2):133–137, 1999.

[51] Schulzrinne H., Casner S., Frederick R. and Jacobson V. RTP: a transport protocol for real-time applications. Available in ftp://ftp.ietf.org/rfc/rfc1889.txt, Jan. 1996.

[52] Schwartz M. *Broadband Integrated Networks*. Prentice Hall, 1996.

[53] ITU–T Recommendation G.729 - Annex B: A silence compression scheme for G.729 optimized for terminals conforming to recommendation v. 70, Nov. 1996.

[54] Sen P., Karlsson G., Maglaris B., Anastassiou D. and Robbins J. Performance models of statistical multiplexing in packet video communications. *IEEE Transactions on Communications*, 36(7):834–844, 1988.

[55] Dai M. and Loguinov D. Analysis and modeling of MPEG-4 and H.264 multi-layer video traffic. In *Proceedings of IEEE Infocom*, pp. 2257–2267, March 2005.

[56] Schwartz M., Skelly P. and Dixit S. A histogram-based model for video traffic behavior in an ATM multiplexer. *IEEE Transactions on Networking*, 1(4):446–459, 1993.

[57] Dawood A.M. and Ghanbar M. Content-based MPEG video traffic modeling. *IEEE Transactions on Multimedia*, 1:77–87, 1999.

[58] Melamed B. and Pendarakis D. Modeling full-length VBR video using Markov-renewal-modulated TES models. *IEEE Journal on Selected Areas in Communication*, 16:600–611, 1998.

[59] Krunz M.M. and Makowski A.M. Modeling video traffic using $M/G/\infty$ input processes: A compromise between Markovian and LRD models. *IEEE Journal on Selected Areas in Communication*, 16:733–748, 1998.

[60] Ramakrishnan S., Sarkar U.K. and Sarkar D. Modeling full-length video using Markov-modulated gamma-based framework. *IEEE Transactions on Networking*, 11(4):638–649, 2003.

[61] Joint Video Team (JVT) of ISO/IEC MPEG and ITU-T VCEG. Draft ITU-T recommendation and final draft international standard of joint video specification (ITU-T Rec. H.264/ISO/IEC 14496-10 AVC), JVT-G050, 2003.

[62] Heyman D.P. and Lakshman T.V. What are the implication of long-range dependence for VBR-video traffic engineering. *IEEE/ACM Transactions on Networking*, 4(3):301–317, 1996.

[63] Shihab E., Wan F., Cai L., *et al.* Performance analysis of IPTV traffic in home networks. In *Proceedings of IEEE Globecom'07*, pp. 5341–5345, Washington, DC, USA, Nov./Dec. 2007.

[64] Wan F., Cai L. and Gulliver A. A simple, two-level Markovian traffic model for IPTV video sources. In *Proceedings of IEEE Globecom'08*, New Orleans, LA, USA, Nov./Dec. 2008.

[65] http://trace.eas.asu.edu/h264/index.html, 2007.

[66] Garrett M.W. and Willinger W. Analysis, modeling and generation of self-similar VBR video traffic. In *Proceedings of ACM SIGCOMM*, pp. 269–280, Aug./Sept. 1994.

[67] Laub A.J. *Matrix Analysis for Scientists and Engineers*. SIAM Publications, 2005.

[68] Chang C.-S. Stability, queue length, and delay, II: Stochastic queueing networks. Report rc 17709, IBM Research, Yorktown Hts., NY, Feb. 1992.

[69] Kesidis G., Walrand J. and Chang C.-S. Effective bandwidths for multiclass Markov fluids and other ATM sources. *IEEE/ACM Transactions on Networking*, 1(4):424–428, 1993.

[70] Floyd S. and Jacobson V. Link-sharing and resource management models for packet networks. *IEEE/ACM Transactions on Networking*, 3(4):365–386, 1995.

[71] Floyd S. and McCanne S. Network Simulator. LBNL public domain software. Available via ftp from ftp.ee.lbl.gov. NS-2 is available at www.isi.edu/nsnam/ns/.

[72] IEEE standard part 15.3: Wireless medium access control (MAC) and physical layer (PHY) specifications for high rate wireless personal area networks (WPANs). IEEE Std 802.15.3-2003, Sept. 2003.

[73] Cai L.X., Cai L., Shen X. and Mark J.W. Capacity of UWB networks supporting multimedia services. In *Proceedings of QShine'06*, Aug. 2006.

[74] Deng S. Empirical model of www document arrivals at access link. In *Proceedings of the IEEE International Conference on Communications (ICC)*, pp. 1797–1802, June 1996.

[75] Tan W. and Zakhor A. Real-time Internet video using error resilient scalable compression and TCP-friendly transport protocol. *IEEE Transactions on Multimedia*, 1(2):172–186, 1999.

[76] Mathis M., Semke J., Mahdavi J. and Ott T. The macroscopic behavior of the TCP congestion avoidance algorithm. *ACM Computer Communication Review*, 27(3):67–82, 1997.

[77] Padhye J., Firoiu V., Towsley D. and Kurose J. Modeling TCP throughput: a simple model and its empirical validation. In *Proceedings of ACM SIGCOMM'98*, pp. 303–314, 1998.

[78] Floyd S., Handley M. and Padhye J. A comparison of equation-based and AIMD congestion control, May 2000. Available http://www.aciri.org/tfrc/tcp-friendly.TR.ps.

[79] Rejaie R., Handley M. and Estrin D. RAP: An end-to-end rate-based congestion control mechanism for realtime streams in the Internet. In *Proceedings of IEEE Infocom'99*, pp. 1337–1345, March 1999.

[80] Karandikar S., Kalyanaraman S., Bagal P. and Packer B. TCP rate control. *ACM Computer Communications Review*, 30(1):45–58, 2000.

[81] Cai L., Shen X., Pan J. and Mark J.W. Performance analysis of TCP-friendly AIMD algorithms for multimedia applications. *IEEE Transactions on Multimedia*, 7(2):339–355, 2005.

[82] Cai L., Shen X., Mark J.W. and Pan J. QoS support for multimedia traffic in wireless/wired networks using TCP-friendly AIMD protocol. *IEEE Transactions on Wireless Communications*, 5(2):469–480, 2006.

[83] Floyd S. and Kohler E. Profile for DCCP congestion control ID 2: TCP-like congestion control, March 2005. Work in progress. Available at www.icir.org/kohler/dcp/draft-ietf-dccp-ccid2-10.txt.

[84] Blanton E., Allman M., Fall K. and Wang L. A conservative selective acknowledgment (SACK)-based loss recovery algorithm for TCP. *IETF RFC 3517*, 2003.

[85] Kelly F.P. Stochastic models of computer communication systems. *Journal of the Royal Statistical Society*, B47(3):379–395, 1985.

[86] Loguinov D. and Radha H. End-to-end rate-based congestion control: Convergence properties and scalability analysis. *IEEE/ACM Transactions on Networking*, 11(4):564–577, 2003.

[87] Bansal D. and Balakrishnan H. TCP-friendly congestion control for real-time streaming applications. *Technical Report MIT-LCS-TR-806*. MIT, May 2000.

[88] Chiu D. and Jain R. Analysis of the increase/decrease algorithms for congestion avoidance in computer networks. *Journal of Computer Networks and ISDN*, 17(1):1–14, 1989.

[89] Gorinsky S. and Vin H. Extended analysis of binary adjustment algorithms. *Technical Report TR2002-39*, University of Texas at Austin, Aug. 2002.

[90] Rejaie R., Handley M. and Estrin D. Quality adaption for congestion controlled video playback over the Internet. In *Proceedings of ACM SIGCOMM'99*, pp. 189–200, 1999.

[91] Zhang Z., Wang Y., Du D.H.C. and Su D. Video staging: a proxy-server-based approach to end-to-end video delivery over wide-area networks. *IEEE/ACM Transactions on Networking*, 8(4):429–442, 2000.

[92] Floyd S. and Kohler E. Internet research needs better models. *ACM SIGCOMM Computer Communication Review*, 33(1):29–34, 2003.

[93] Floyd S. and Jacobson V. On traffic phase effects in packet-switched gateways. *Internetworking: Research and Experience*, 3(3):115–156, 1992.

[94] Jain R. *The Art of Computer Systems Performance Analysis: Techniques for Experimental Design, Measurement, Simulation, and Modeling*. John Wiley & Sons, New York, NY, April 1991.

[95] Ramjee R., Kurose J.F., Towsley D.F. and Schulzrinne H. Adaptive playout mechanisms for packetized audio applications in wide-area networks. In *Proceedings of IEEE Infocom'94*, pp. 680–688, 1994.

[96] Khalil H.K. *Nonlinear Systems*. Prentice Hall, Upper Saddle River, N.J., 2002.

[97] Luenberger D.G. *Introduction to Dynamic Systems: Theory, Models, and Application.* John Wiley & Sons, May 1979. ISBN: 0-471-02594-1.

[98] Low S., Paganini F., Wang J., *et al.* Dynamics of TCP/RED and a scalable control. In *Proceedings of IEEE Infocom'02*, vol. 1, pp. 239–248, June 2002.

[99] Wang L., Cai L., Liu X. and Shen X. Stability and TCP-friendliness of AIMD/RED systems with feedback delays. *Elsevier Journal of Computer Networks*, 51(15):4475–4491, 2007.

[100] Roy R., Raghuraman M. and Panwar S.S. Analysis of TCP congestion control using a fluid model. In *Proceedings of IEEE ICC'01*, pp. 2396–2403, 2001.

[101] Misra V., Gong W.B. and Towsley D. Fluid-based analysis of a network of AQM routers supporting TCP flows. In *Proceedings of ACM/SIGCOMM*, pp. 151–160, 2000.

[102] Hollot C.V., Towsley D., Misra V. and Gong W.B. Analysis and design of controllers for AQM routers supporting TCP flows. *IEEE Transactions on Automatic Control*, 47(6):945–959, 2002.

[103] Hollot C.V. and Chait Y. Non-linear stability analysis of a class of TCP/AQM networks. In *Proceedings of IEEE Conference on Decision and Control*, vol. 3, pp. 2309–2314, Orlando, FA, USA, Dec. 2001.

[104] Liu S., Basar T. and Srikant R. Pitfalls in the fluid modeling of RTT variations in window-based congestion control. In *Proceedings of IEEE Infocom'05*, vol. 2, pp. 1002–1012, Miami, FA, USA, Mar. 2005.

[105] Wang L., Cai L., Liu X. and Shen X. Practical stability and bounds of heterogeneous AIMD/RED system with time delay. In *Proceedings of IEEE ICC'08*, pp. 5558–5563, Beijing, China, May 2008.

[106] Fairhurst G. and Wood L. Advice to link designers on link Automatic Repeat reQuest (ARQ). *IETF RFC 3366*, 2002.

[107] Esteves E. The high data rate evolution of the CDMA2000 cellular system. *Multiaccess, Mobility and Teletraffic for Wireless Communications*, 5:61–72, 2000.

[108] Zhang D. and Wasserman K.M. Transmission schemes for time-varying wireless channels with partial state observations. In *Proceedings of IEEE Infocom'02*, vol. 2, pp. 467–476, 2002.

[109] Bhagwat P., Bhattacharya P.P., Krishna A. and Tripathi S.K. Enhancing throughput over wireless LANs using channel state dependent packet scheduling. In *Proceedings of IEEE Infocom'96*, pp. 1133–1140, 1996.

[110] Chan M.C. and Ramjee R. TCP/IP performance over 3G wireless links with rate and delay variation. In *Proceedings of ACM Mobicom'02*, pp. 71–82, Sept. 2002.

[111] Chockalingam A., Zorzi M. and Rao R.R. Performance of TCP on wireless fading links with memory. In *Proceedings of IEEE ICC'98*, 1998.

[112] Zorzi M. Data-link packet dropping models for wireless local communications. *IEEE Transactions on Vehicular Technology*, 51(4):710–719, 2002.

[113] Zorzi M. and Rao R.R. ARQ error control for delay-constrained communications on short-range burst-error channels. In *Proceedings of IEEE VTC'97*, May 1997.

[114] Casetti C. and Meo M. A new approach to model the stationary behavior of TCP connections. In *Proceedings of IEEE Infocom'00*, vol. 1, pp. 367–375, 2000.

[115] Kherani A.A. and Kumar A. Closed loop analysis of the bottleneck buffer under adaptive window controlled transfer of HTTP-like traffic elastic flows in the Internet. In *Proceedings of IEEE Infocom'03*, vol. 2, pp. 906–915, April 2003.

[116] Fendick K.W., Rodrigues M.A. and Weiss A. Analysis of a rate-based feedback control strategy for long haul data transport. *Performance Evaluation*, 16:67–84, 1992.

[117] Fendick K.W., Rodrigues M.A. and Weiss A. Analysis of a rate-based control strategy with delayed feedback. *ACM SIGCOMM Computer Communication Review*, 2(4):136–148, 1992.

[118] Cai L., Shen X., Mark J.W. and Pan J. Performance modeling and analysis of window-controlled multimedia flows in wireless/wired networks. *IEEE Transactions on Wireless Communications*, 6(4):1356–1365, 2007.

[119] Chaskar H.M., Lakshman T.V. and Madhow U. TCP over wireless with link level error control: analysis and design methodology. *IEEE/ACM Transactions on Networking*, 7(5):605–615, 1999.

[120] Ramakrishnan K. and Floyd S. A proposal to add Explicit Congestion Notification (ECN) to IP. *IETF RFC 2481*, Jan. 1999.

[121] Balakrishnan H., Padmanabhan V., Seshan S. and Katz R.H. A comparison of mechanisms for improving TCP performance over wireless links. *IEEE/ACM Transactions on Networking*, 5(6):756–769, 1997.

[122] Balakrishnan H., Seshan S., Amir E. and Katz R.H. Improving TCP/IP performance over wireless networks. In *Proceedings of ACM Mobicom'95*, pp. 2–11, 1995.

[123] Bakre A. and Badrinath B.R. I-TCP: Indirect TCP for mobile hosts. In *Proceedings of ICDCS'95*, pp. 136–143, 1995.

[124] Brown K. and Singh S. M-TCP: TCP for mobile cellular networks. *ACM Computer Communication Review*, 27(5):19–43, 1997.

[125] Balakrishnan H., Seshan S. and Katz R.H. Improving reliable transport and handoff performance in cellular wireless networks. *ACM Wireless Networks*, 1(4):469–481, 1995.

[126] Chan A., Tsang D.H.K. and Gupta S. Impact of handoff on TCP performance in mobile wireless computing. In *Proceedings of IEEE ICPWC'97*, pp. 184–188, 1997.

[127] Goff T., Moronski J., Phatak D.S. and Gupta V. Freeze-TCP: A true end-to-end enhancement mechanism for mobile environments. In *Proceedings of IEEE Infocom'00*, pp. 1537–1545, 2000.

[128] Pan J. TCP/IP over air links, literature scanning. Available at http://bbcr.uwaterloo.ca/~jpan/tcpair/, March 2001.

[129] Semke J., Mahdavi J. and Mathis M. Automatic TCP buffer tuning. In *Proceedings of ACM SIGCOMM'98*, pp. 315–323, 1998.

[130] Kalampoukas L., Varma A. and Ramakrishnan K.K. Explicit window adaptation: a method to enhance TCP performance. *IEEE/ACM Transactions on Networking*, 10(3):338–350, 2002.

[131] Chan M.C. and Ramjee R. Improving TCP/IP performance over third generation wireless networks. In *Proceedings of IEEE Infocom'04*, 2004.

[132] Zorzi M., Rao R.R. and Milstein L.B. ARQ error control for fading mobile radio channels. *IEEE Transactions on Vehicular Technology*, 46(2):445–455, 1997.

[133] Badrinath B.R. and Sudame P. To send or not to send: implementing deferred transmissions in a mobile host. In *Proceedings of ICDCS'96*, pp. 327–333, 1996.

[134] Vojnovic M., Boudec J.L. and Boutremans C. Global fairness of Additive-Increase and Multiplicative-Decrease with heterogeneous round-trip times. In *Proceedings of IEEE Infocom'00*, vol. 3, pp. 1303–1312, 2000.

[135] Gurtov A., Passoja M., Aalto O. and Raitola M. Multi-layer protocol tracing in a GPRS network. In *Proceedings of IEEE VTC'02*, vol. 3, pp. 1612–1618, 2002.

[136] Ciavattone L., Morton A. and Ramachandran G. Standardized active measurements on a tier 1 IP backbone. *IEEE Communications*, June 2003. Available at http://ipnetwork.bgtmo.ip.att.net/pws/att_ieee.pdf.

[137] The PingER project, 2004. Available at www-iepm.slac.stanford.edu/pinger/.

[138] Packet trace analysis, 2004. Available at http://ipmon.sprint.com/ipmon.php.

[139] MCI service level agreements. Available at http://global.mci.com/terms/us/products/internet/sla/.

[140] Sprint service level agreements. Available at http://www.sprint.com/business/support/serviceLevelAgreements.jsp.

[141] AT&T service level agreements. Available at http://www.business.att.com.

[142] Braden R. Requirements for Internet hosts — communication layers. *IETF RFC 1122*, Oct. 1989.

[143] Floyd S., Handle M., Padhye J. and Widmer J. Equation-based congestion control for unicast applications. In *Proceedings of ACM SIGCOMM'2000*, pp. 43–56, 2000.

[144] Padhye J., Firoiu V., Towsley D. and Kurose J. Modeling TCP throughput: a simple model and its empirical validation. In *Proceedings of ACM SIGCOMM'98*, pp. 303–314, 1998.

[145] Handley M., Floyd S., Padhye J. and Widmer J. TCP Friendly Rate Control (TFRC): Protocol specification. *IETF RFC 3448*, Jan. 2003.

[146] Shen H., Cai L. and Shen X. Performance analysis of TFRC over wireless links with link level ARQ. *IEEE Transactions on Wireless Communications*, 5(6):1479–1487, 2006.

[147] Crow B.P., Widjaja I., Kim J.G. and Sakai P.T. Investigation of the IEEE 802.11 medium access control (MAC) sublayer functions. In *Proceedings of IEEE Infocom'97*, vol. 13, pp. 126–133, April 1997.

[148] Veeraraphavan M., Cocker N. and Moors T. Support of voice services in IEEE 802.11 wireless LANs. In *Proceedings of IEEE Infocom'01*, vol. 1, pp. 448–497, Anchorage, Alaska, April 2001.

[149] Xiao Y. IEEE 802.11e: QoS provisioning at the MAC layer. *Wireless Communications*, 11(3):72–79, 2004.

[150] IEEE 802.11 WG. Part 11: Wireless LAN Medium Access Control (MAC) and Physical Layer (PHY) specification, Aug. 1999.

[151] Bianchi G. Performance analysis of the IEEE 802.11 distributed coordination function. *IEEE Journal on Selected Areas in Communication*, 18(3):535–547, 2000.

[152] Cai L.X., Shen X., Mark J.W., Cai L. and Xiao Y. Voice capacity analysis of WLAN with un-balanced traffic. *IEEE Transactions on Vehicular Technology*, 55(3):752–761, 2006.

[153] Zhai H., Chen X. and Fang Y. How well can the IEEE 802.11 wireless LAN support quality of service. *IEEE Transactions on Wireless Communications*, 4(6):3084–3094, 2005.

[154] Tickoo O. and Sikdar B. Queueing analysis and delay mitigation in IEEE 802.11 random access MAC based wireless networks. In *Proceedings of IEEE Infocom'04*, vol. 2, pp. 1404–1413, Hong Kong, China, Mar. 2004.

[155] Tickoo O. and Sikdar B. A queueing model for finite load IEEE 802.11 random access MAC. In *Proceedings of IEEE ICC'04*, vol. 1, pp. 175–179, June 2004.

[156] Heck, A. *Introduction to Maple (3rd edn)*. Springer-Verlag, New York, 2003.

[157] Hole D.P. and Tobagi F.A. Capacity of an IEEE 802.11b wireless LAN supporting VoIP. In *Proceedings of IEEE ICC'04*, vol. 1, pp. 196–201, June 2004.

[158] Garg S. and Kappes M. Can I add a VoIP call? In *Proceedings of IEEE ICC'03*, vol. 2, pp. 779–783, May 2003.

[159] ITU-T Recommendation G. 114. One-way transmission time, Series G: Transmission systems and media, digital systems and networks, International telephone connections and circuits – General recommendations on the transmission quality for an entire international telephone connection, May 2003.

[160] Ott J. and Kutscher D. Drive-thru Internet: IEEE 802.11b for automobile users. In *Proceedings of IEEE Infocom'04*, Mar. 2004.

[161] Mancuso V. and Bianchi G. Streaming for vehicular users via elastic proxy buffer managment. *IEEE Communication Magazine*, 42(11):144–152, 2004.

[162] Ho D. and Valaee S. Information raining and optimal link-layer design for mobile hotspots. *IEEE Transactions on Mobile Computing*, 4(3):271–284, 2005.

[163] Wimax forum: http://www.wimaxforum.org/home/.

[164] Pack S., Shen X., Mark J.W. and Cai L. Throughput analysis of TCP friendly rate control in mobile hotspots. *IEEE Transactions on Wireless Communications*, 7(1):193–203, 2008.

[165] Kim J., Kim S., Choi S. and Qiao D. CARA: Collision-aware rate adapatation for IEEE 802.11 WLANs. In *Proceedings of IEEE Infocom 2006*, Apr. 2006.

[166] Zhai H., Kwon Y. and Fang Y. Performance analysis of IEEE 802.11 MAC protocols in wireless LANs. *Wireless Communications and Mobile Computing*, 4(8):917–931, 2004.

[167] Wu H., Peng Y., Long K., *et al*. Performance of reliable transport protocol over IEEE 802.11 wireless LAN: Analysis and enhancement. In *Proceedings of IEEE Infocom'02*, June 2002.

[168] Chen M. and Zakhor A. Rate control for video streaming over wireless. In *Proceedings of IEEE Infocom'04*, Mar. 2004.

[169] Part 11 IEEE 802.11a. Wireless LAN, medium access control (MAC) and physical layer (PHY) specifications: High-speed physical layer in the 5GHz band, Sept. 1999.

[170] Jacobson V. and Karels M. Congestion avoidance and control. In *Proceedings of ACM SIGCOMM'88*, pp. 314–329, 1988.

[171] Floyd S. and Henderson T. The NewReno modification to TCP's fast recovery algorithm. *IETF RFC 2582*, April 1999.

[172] Mathis M., Mahdavi J., Floyd S. and Romanow A. TCP Selective Acknowledgment option. *IETF RFC 2018*, April 1996.

[173] Braden B., Clark D., Crowcroft J., *et al.* Recommendation on queue management and congestion avoidance in the Internet. *IETF RFC 2309*, April 1998.

[174] Kohler E., Handley M. and Floyd S. Datagram Congestion Control Protocol (DCCP). Work in progress, Dec. 2005. Available at www.icir.org/kohler/dcp/draft-ietf-dccp-spec-13.txt.

[175] Evlogimenos A., Lim K.H. and Lai K. On the implementation of Datagram Congestion Control Protocol, 2002. Available at www.cs.berkeley.edu/~laik/projects/dccp/index.html.

[176] Floyd S., Kohler E. and Padhye J. Profile for DCCP congestion control ID 3: TFRC congestion control, March 2005. Work in progress. Available at www.icir.org/kohler/dcp/draft-ietf-dccp-ccid3-11.txt.

Index

A-to-D conversion, 46
acknowledgment self-clocking property, 144
adaptive modulation and coding (AMC), 143
Additive Increase and Multiplicative Decrease
 (AIMD) congestion control, 76
 convergent speed, 91
 effectiveness, 87
 practical implications, 93
 responsiveness, 90
 TCP-friendly AIMD (TAIMD), 86
AIMD protocol
 acknowledgment, 77
 congestion avoidance, 79
 congestion window ($cwnd$), 79
 dynamic TCP-friendly AIMD
 (DTAIMD) algorithm, 94
 exponential backoff, 79
 flow and congestion control, 78
 implementation concerns, 98
 increase rate and decrease ratio (α, β)
 pair, 81
 receiver window ($rwnd$), 78, 163
 sender window, 79
 slow start, 79
angle-of-arrival (AOA), 25, 27
angular power spectral density (APSD), 25,
 28
auto-covariance function (ACF), 50
Automatic Repeat reQuest (ARQ), 143
 truncated ARQ, 184
average fade duration (AFD), 15
average loss interval method, 181

bandwidth over-provisioning, 160
binary search, 188
binomial congestion control, 86
 inverse increase and additive decrease
 (IIAD), 87
 square-root increase and square-root
 decrease (SQRT), 87

bit error rate (BER)
 Rayleigh fading channel, 11

channel profile, 148
channel state dependent (CSD) transmission,
 143, 148
class-based queueing (CBQ), 60
coherence bandwidth, 16
constant bit rate (CBR), 46, 178
correlation coefficient, 54
correspondent host (CH), 147

delay control, 158
delay outage, 154, 167, 184
 delay outage probability, 158
 delay outage rate, 146
discrete-time Markov chain (DTMC), 182
Doppler frequency shift, 12
Drop-Tail queue, 101

equal probability method (EPM), 18
equilibrium point, 119
Explicit Congestion Notification (ECN), 145
Explicit Loss Notification (ELN), 145
exponentially weighted moving average
 (EWMA) method, 182

fading, *see wireless channel*, 9
fairness index, 100
finite-state Markov chain (FSMC), 12
 for people-shadowing effect, 26
 for video traffic, 50
 frequency-selective channels, 17
 OFDM system, 19
 Rayleigh fading channels, 10
 two-level Markovian model for video,
 51
first-in–first-out (FIFO) queue, 56

fluid-flow model
        for AIMD/RED system, 117

goodput, 178
Group of Picture (GoP), 50, 51

hidden terminal problem, 205

I-TCP, 146
International Telecommunication
        Union-Telecommunication
        (ITU–T), 47
Internet Engineering Task Force (IETF), 47
internet Low Bit Rate Codec (iLBC), 46
Internet Protocol (IP), 1
Internet Protocol Television (IPTV), 3
intersymbol interference (ISI), 14

Kronecker product, 60
Kronecker sum, 60

level-crossing rate (LCR), 15
        in the frequency domain, 16
        Rayleigh fading channel, 11
line-of-sight (LOS), 24
link budget, 34
link throughput distribution, 148
link utilization, 170
loss event, 181
lower bound (LB) on window size, 129
Lyapunov function, 119

M-TCP, 146
Markov chain, 12
        convergent speed, 12
        memory, 12, 13
        state transition probability
           Rayleigh fading channel, 11
        state transition probability, 12
        steady state probability, 11, 12
Mean Opinion Score (MOS), 47
minimum mean square error (MMSE)
        algorithm, 50
mobile host (MH), 146, 182
Moving Picture Experts Group (MPEG), 50
        MPEG-4, 51
multimedia playback application, 95
multimedia traffic, 45

normalized throughput, 99
normalized wireless link utilization, 184

on/off model, 47
Orthogonal Frequency-Division Multiplexing
        (OFDM), 13
        cyclic prefix (CP), 20
        OFDM channel, 14

packet error rate (PER), 35
        Rayleigh fading channel, 11
        UWB channel, 34
packet loss rate (PLR), 51
phase effect, 99
power delay profile (PDP), 15

random early detection (RED), 85, 118
Rate Adaptation Protocol (RAP), 76
Real-Time Transport Control Protocol
        (RTCP), 46
Real-Time Transport Protocol (RTP), 46

scalable source coding, 70
        Fine Granularity Scalability (FGS)
           coding, 70
        Multiple Description (MD) coding, 71
service differentiation, 97, 124
Session Initiation Protocol (SIP), 47
signal-to-noise ratio (SNR), 10
snoop TCP, 146
stability, 115
        asymptotically stable, 116
        marginally stable, 116
        stable, 116
        uniformly bounded, 116
survivor function, 58

TCP-friendliness, 75, 123, 163, 170
TCP-Friendly Rate Control (TFRC) protocol,
        181
token control, 152
traffic model, 45
Transmission Control Protocol (TCP), 75

ultra-wideband (UWB), 24
        channel model, 26
        link budget, 34
        multiband OFDM (MB-OFDM), 33
        people-shadowing effect, 31
upper bound (UB) on queue length, 129
upper bound (UB) on window size, 128
User Datagram Protocol (UDP), 5

variable bit rate (VBR), 51
video performance study, non-adaptive video
     sources, 55
   heterogeneous traffic with class-based
     queueing, 60
   over wired link, 58
   over wireless link, 59
video traffic model
   two-level Markovian model, 51
   mini-source model, 49
   motion complexity, 52
   spatial correlation, 52
   temporal correlation, 52
   texture complexity, 52
Voice over IP (VoIP), 45
voice traffic model, 47
   on/off model, 47

wireless channel, 9
   frequency-nonselective (flat) fading, 10
   frequency-selective fading, 14
   large-scale fading, 9

multipath fading, 9
Nakagami-*m* fading, 15, 16
path loss, 9
Rayleigh fading, 10
shadowing, 25
small-scale fading, 9
wireless channel model, 9
   for frequency-selective fading channels,
     13
   for indoor wireless channels with
     shadowing, 24
   for Rayleigh fading channels, 10
   Jakes' model, 15
wireless channel profile, 148
Wireless Local Area Network (WLAN), 202
   Distributed Coordination Function
     (DCF), 204
   infrastructure-based, 210
   Point Coordination Function (PCF), 202
   ready-to-send/clear-to-send (RTS/CTS),
     204